"Insightful, lyrical . . . *Little and Often* proves to be a rich tale of self-discovery and reconciliation. Resonating with Robert Pirsig's classic *Zen and the Art of Motorcycle Maintenance*, it is a profound father-and-son odyssey."

—***USA TODAY*** **(four stars)**

"An impressive memoir and a richly rendered tale. I may have grown misty at one point."

—**NICK OFFERMAN**, **actor and *New York Times* bestselling author**

"The woodworking is rich and beyond impressive, but Preszler's humble soul work is utterly transcendent."

—**MATTHEW QUICK**, ***New York Times*** **bestselling author of *The Silver Linings Playbook* and *The Reason You're Alive***

"Masterful. With *Little and Often*, Trent Preszler gives us not only a memoir of sons and fathers, acceptance and reconciliation, but also a stirring meditation on objects, their memories, and the complexities of inheritance."

—**GRANT GINDER**, **author of *The People We Hate at the Wedding* and *Honestly, We Meant Well***

"Woodworking meets bridge-building, and sorrow meets understanding in this impeccably written, loving memoir."

—***KIRKUS REVIEWS*** **(starred review)**

"In Trent Preszler's hands, we are smoothed, soothed, and made anew as he peels back layer after layer of his grief and loss until there is only love and forgiveness. This is an unforgettable story of a father's final, life-altering gift to his son."

—**DANI SHAPIRO**, ***New York Times*** **bestselling author of *Inheritance***

Praise for

Little and Often

"Insightful, lyrical . . . *Little and Often* proves to be a rich tale of self-discovery and reconciliation. Resonating with Robert Pirsig's classic *Zen and the Art of Motorcycle Maintenance*, it is a profound father-and-son odyssey that discovers the importance of the beauty of imperfection and small triumphs that make extraordinary happen."

—*USA Today* (four stars)

"The woodworking is rich and beyond impressive, but Preszler's humble soul work is utterly transcendent. Courageous. Genuine. Cathartic. Will restore your faith in forgiveness. Will make you believe in grace."

—Matthew Quick, *New York Times* bestselling author of *The Silver Linings Playbook* and *The Reason You're Alive*

"An impressive memoir and a richly rendered tale. I thought (with relish) that I was getting a book about wood and tools, but the canoe built herein is merely the vessel carrying the buoyant narrative about a father and son, a mother and sister, love, hard work, wine, boats, and a dog. I may have grown misty at one point."

—Nick Offerman, actor and *New York Times* bestselling author

"*Little and Often* is a beautiful memoir of grief, love, the shattered bond between a father and son, and the resurrection of a broken heart. Trent Preszler tells his story with the same level of art and craftsmanship that he brings to his boat making, and he reminds us of creativity's power to transform and heal our lives. This is a powerful and deeply moving book. I won't soon forget it."

—Elizabeth Gilbert

"Sometimes a writer goes on a journey in order to write a book. More rarely, a writer writes a book in order to go on a journey. *Little and Often* belongs to that latter category of memoir, built from the inside out. In Trent Preszler's hands, we are smoothed, soothed, and made anew as he peels back layer after layer of his grief and loss until there is only love and forgiveness. This is an unforgettable story of a father's final, life-altering gift to his son."

—Dani Shapiro, *New York Times* bestselling author of *Inheritance*

"Trent Preszler's beautiful, compelling memoir tells of his struggle that spans a divided country and family alike. The writing is simple and elegant, harkening back to great American writers such as John Edward Williams and Willa Cather. The expanse between South Dakota, New York City, and, finally, the North Fork of Long Island is enormous for a young gay man struggling with his father's legacy. It's a tenderly wrought tale of coming to terms with your past that will resonate no matter who you are."

—Isaac Mizrahi, fashion designer and host of *Project Runway*

"Woodworking meets bridge-building, and sorrow meets understanding in this impeccably written, loving memoir."

—*Kirkus Reviews* (starred review)

"Masterful. With *Little and Often*, Trent Preszler gives us not only a memoir of sons and fathers, acceptance and reconciliation, but also a stirring meditation on objects, their memories, and the complexities of inheritance. The prose is crystalline, and Preszler's voice is as sure as the steadiest canoe."

—Grant Ginder, author of *The People We Hate at the Wedding* and *Honestly, We Meant Well*

"*Little and Often* is a meditation on spiritual growth, nature's magic, the love for family, regret, and the redemptive power of craftsmanship. I have the highest regard for Trent's courage in writing this big and beautiful memoir. It's a soulful and sometimes gut-wrenching story of the difficult relationships between fathers and sons. This gem couldn't be more relevant to the times we live in today."

—Kevin O'Connor, host of *This Old House* on PBS

"Ultimately it's a tale as well-crafted as the beautiful canoe."

—*Booklist*

"*Little and Often* is filled with joy."

—*Denver Post*

LITTLE AND OFTEN

LITTLE AND OFTEN

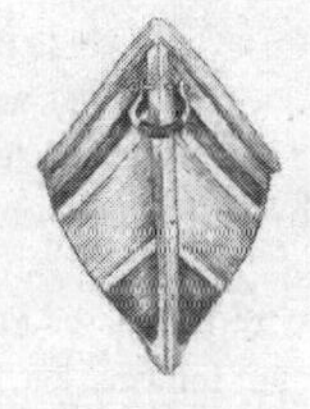

A MEMOIR

TRENT PRESZLER

wm

WILLIAM MORROW

An Imprint of HarperCollins*Publishers*

A hardcover edition of this book was published in 2021 by William Morrow, an imprint of HarperCollins Publishers.

FIRST WILLIAM MORROW PAPERBACK EDITION PUBLISHED 2022.

Designed by Leah Carlson-Stanisic

Library of Congress Cataloging-in-Publication Data has been applied for.

ISBN 978-0-06-297665-9

22 23 24 25 26 LSC 10 9 8 7 6 5 4 3 2 1

In memory of Leon K. Preszler and Lucinda M. Preszler

Dedicated to Christine S. Hoover, with love and gratitude

And for my mother, who taught me something about survival

CONTENTS

LITTLE AND OFTEN

AUTHOR'S NOTE

I have made my best efforts to ensure accuracy of detail and emotion in this recounting of my life. I changed the names and identifying personal details of certain people who appear in the book to preserve their anonymity. As my memory is sometimes fallible, there are places in the text where some dialogue is approximated, combined, or moved in time. Digressive conversations have been shortened. I omitted specific people and events—intentionally, or because of compressed event timelines—but only when those omissions had no impact on the substance of the story. There has been editing, but not at the cost of essential truthfulness.

The imposition of structure to any story, I have learned, alters realities. Scenes plucked from the fabric of life inevitably change when contemplated years after their native context. Nuances of character are sometimes lost or enhanced to the considerations of narrative. Still, this is the story of my life, created from my memories and perceptions of events, filtered through time. Others may have their own thoughts and interpretations. My greatest wish in writing this book is to help people, and I believe we are most inspiring when revealed at our most human.

A man can be destroyed but not defeated.

Ernest Hemingway,
The Old Man and the Sea

Where no sea runs, the waters of the heart
Push in their tides

Dylan Thomas,
"Light Breaks Where No Sun Shines"

PROLOGUE

First there was cedar. Then there was walnut. Then, in my house above Peconic Bay at dusk, I was ready to carve the ash. Every time I twisted my hands to ease off the cuts, curled shavings fluttered silently to the floor. The nutty scent of the wood hung in the air. A chaos of shavings piled up like snow. In that form, they were incomprehensible and unpromising. It was hard to believe something so beautiful had grown from such disarray.

Outside, a storm squall moved in from the northeast, whipping up whitecaps on the sea. Everything was gray. The last few yellow leaves clung to the scrub oaks on the beach. Had it already been a year?

I had left off the night before at the bow, where the gunwales met in a point. There was one side still to finish, and then a lot of difficult sanding after that. The top edge must gradually curl and blend into the hull, so hardwood and softwood become part of the same aerodynamic shape, the same current of energy. My eyes stayed on the top railing, and out went my right hand to land on the spokeshave. Long ago, I had memorized its placement, and my hand travelled there instinctively. Despite my struggles with them early on, by now the tools had become familiar through repetition. They revealed themselves to me by their weight and balance, by the texture of their cold metal surfaces, by the feel of their soft, worn handles.

I leaned against the hull; its thin planking pressed hard on my hip. Now the stroke, long and with the grain, walking the length of the boat as the blade zipped through the tree's growth rings, its cambium, sapwood, heartwood, and pith. The cut was crisp, with little resistance, like the feel and sound when you turn the crank of a pencil sharpener. The gunwales were milled from a rare beetle-free log of American white ash, and the tree must have been old when it fell to the logger's saw. Well, old for a person, young for a tree.

My whole body propelled the tool, reaching forward and pulling back. Each stroke of the spokeshave required a slightly new way to grip it, a slightly different arm and shoulder position, a little compensation left or right with my torso. It was a pleasing twist, like kayaking, each stroke different from all the others before it, and each shaving a unique imprint of the blade. To travel that long distance between rough and smooth, I had to trust the process: remove a little bit here, a little more there, until all the little cuts added up to something bigger.

Ducks passed by on their southern migration. Water ripped through the channel on its path to the sea. The clouds darkened as the squall closed in. It didn't matter; there was nowhere else I would rather be.

I moved deeper into the carving, working silently, no words or thoughts to distract me. It felt timeless. A trancelike hum bled away whatever anxieties I still harbored. My ears tuned in to the music of sharpened steel slicing through wood, and the wood was saying things to me—things I had only recently begun to understand.

CHAPTER 1

THE PHONE CALL

"Hello, Trent?"

The two words froze me. During the fourteen years my father and I were estranged, he called only when he had devastating news to share. His voice transported me to the last time I had answered a phone and heard him say my name. I thought of that day, and of what he had said next. I hoped he was calling for a different reason this time.

"Mom told me you was livin' on the ocean now," he said.

I kept up with family news through a back channel: Mom. We would chat on her desk line at the University of South Dakota, where she worked as a support specialist. It had not occurred to me that she was telling my father everything I had told her over the years.

"Yes, I moved into a beach house."

A divorce earlier that year, in 2014, had left me homeless at age thirty-seven. I was living out of a suitcase in a hotel room when I read an ad in the local paper for a 1920s beach bungalow. I liked it so much that I signed a lease on the spot. My new house looked across a bay toward Robins Island, a rocky outcropping that was deposited as glacial outwash eleven thousand years ago. The channel between my house and the island was deep, and during the changing of the tides, it filled with clear seawater that flowed swiftly past Marratooka Point. The house was perched on a bluff eighty miles east of New York City in a landscape that I had come to love—a place where Long Island's suburban strip malls gave way to bleached sand dunes. It was a land of pitch pine and scrub oak barrens, manicured vineyards, polo matches, and roadside farm stands selling heirloom vegetables to weekenders from the city. In summer, there was constant traffic on Main Road and private jets roaring to the Hamptons. The culture and geography of this place revolved around the Peconic Estuary system: more than a hundred distinct bays, harbors, and creeks that ebbed and flowed in a relentless churn, mixing the waters of Long Island Sound with the Atlantic Ocean. Nothing on the beach was ever the same from one day or even one hour to the next. Some days the bay in front of my house was wild and violent, with thrashing four-foot waves. Other days, the water was flat and calm like an alpine lake, a study of infinite shades of blue.

"That's awful big water out there," my father said. "You got a boat?"

He spoke in a firm kind of cowboy poetry, made of actions and things. This time, his unexpectedly weak and gravelly voice distracted me from the substance of his question. At some point, the strong voice I once knew as my father's must have turned into this old man's voice I heard on the phone. I did not know when the change happened. Perhaps there was no single moment.

"No boat—don't have time," I said.

"How's work?"

"What?" I was so surprised to be asked anything about my personal life that I became confused, as if I had been reading something and lost my place on the page. "Oh, work is . . . fine."

After a decade cutting my teeth in the wine business, I had recently been promoted to CEO of a vineyard on the North Fork of Long Island. My new title was not a respite, just more pressure. I had not taken a day off in months and every minute was booked with board meetings, wine tastings, charity galas, and lavish client dinners at restaurants that were almost impossible to get into. On top of that, I received daily reminders from the owner that the winery had better turn a profit soon, or else my stint as CEO would be short. The workload was brutal, but mine, and better than sitting in a high-rise cubicle. The day my father called I was making final preparations to unveil our new wine label at a launch party in Manhattan the week before Thanksgiving.

A few moments passed with only the sound of my father's labored breathing on the phone. Mom's voice came from somewhere in the background, muffled: "Well, didya ask him?"

He cleared his throat. "We was hopin' you'd come home for Thanksgiving this year."

There it was: an invitation back into my father's life in South Dakota. I couldn't remember the last time I had spent a holiday back home. I had given up hope of reconciliation years ago and accepted our relationship for what it was. I became suddenly nervous, as if by agreeing to see him for the first time in a long while—for the altogether normal purpose of sharing Thanksgiving dinner—I would cross a line and put myself at risk.

"I wantya to know," my father continued, his voice trailing away from the receiver, "that Mom thinks this is a good idea."

He waited for an answer while I stood in the doorway of my house, sweating, my phone pressed against my right ear, staring at the whitecaps on the bay. I had to think fast. I fidgeted with my dog Caper's leash around my wrist. Before I spoke, I swallowed but my mouth was dry.

"Sorry, I have to work."

I was wrenched by guilt as soon as I said it. For years my father and I had been locked in a silent conflict of wills, but in that moment,

I realized how much I wanted that conflict to end, and how deeply I hoped for a future where we could be a father and son at peace.

He coughed—a feeble, rasping hack. Too much time had passed. I had to say something. If my job was not a good enough excuse, I threw out another one that he might relate to.

"Besides, I just adopted a puppy and can't leave him here."

"Bring the dog and bring a bottle of yer fancy wine."

CHAPTER 2

FIRST CRUSH

The tension in the grill room at the Four Seasons Restaurant was palpable. I had been reading about our launch party in the papers for the past week, ever since the story broke in *T: The New York Times Style Magazine*. The articles were saying it was going to be a big night, headlined by the artist I had commissioned to design our wine label. She was a rising star in the New York art world, well-known for her portrait of First Lady Michelle Obama, which had been acquired by the Smithsonian in 2010. Tonight's event showcased the first new wine release I had overseen since becoming CEO, and I could hardly contain my nerves.

The room shimmered with gold chainmail curtains and crystal candelabras. Bundles of pink cherry tree branches—forced into bloom in late November—looked as if they grew from the tables themselves. Bone white china rested on crisp linen tablecloths framed by red suede banquets. The whole scene was staged deliberately around a display table in the center of the room, where a single bottle glowed under a spotlight. It was our new wine, called *First Crush*, and the label bore a joyful painting of a woman's face that was obscured by strips of red, yellow, and purple paper. The collage created a three-dimensional illusion: there was the person obvious at first glance, and the person that existed below the surface. I would not dare tell anyone my real opinion of the art, but secretly, the Picasso-like portrait with the woman's crooked red lipstick was too garish for my taste.

My boss, the winery owner Michael Lynne, arrived wearing a tailored suit and flanked by assistants. He was more famous for his day job as the recently retired CEO of New Line Cinema and executive producer of *The Lord of the Rings* trilogy. He was not handsome in the typical way—his hair too thin and face too hard—but he commanded the room with his confidence.

My phone buzzed, and I pulled it out of my pocket to see a text from Mom: "Did you decide if you're coming home for Thanksgiving? Dad wants to know."

Before I could reply, Michael interrupted.

"We're the hottest ticket in town tonight," he said, with a hug and a slap on the back. "And you made everything happen, Trent—you did it!"

"Thanks, I'll be relieved when it's over."

"What's wrong? You look like you've seen a ghost."

"Nothing," I said, turning off my phone and putting it away.

An editor from the *New York Post* was the first guest to arrive—a woman of unflappable poise and elegance, wearing a look straight out of New York Fashion Week.

"Good evening," she said while lifting a glass of First Crush from a silver platter balanced on the outstretched arm of a tuxedo-clad waiter. Next came the director of the Studio Museum in Harlem and an NBC News anchorman, a supermodel in an orange dress, and a steady stream of Wall Street moguls and Chelsea gallerists who mingled with magazine editors, hip-hop stars, and blue-chip artists while noshing on caviar blintzes. Lights shimmered off the gold curtains in rhythm with the music a DJ spun from the mezzanine. Waiters hurried out of the kitchen carrying sizzling platters of filet mignon and lobster. The aromas of perfume and sweat, garlic and wine permeated the room. I recognized names, posed for photos, and managed to avoid saying anything embarrassing.

Michael was in his element. I tried to emulate his speech, his mannerisms, and the way he carried himself in the room. He always led with "how nice to see you," so that the person thought he remembered them from another time and place, even if they were a total stranger. The guests exclaimed about the flowers, the art, the wine, and how lovely it was to be invited. Was it as easy for Michael as it looked? To move so lightly, to begin sentences by saying with a smile, "Tell me, how was Cannes this year?" and for the person to smile back and say, "It was marvelous" or "It was ghastly." And "Tell me about your favorite wine," and the replies were either "Yes, I know all about wine" or "You know, I don't know the first thing about wine." It did not matter which, either one was an amusing answer in a New York culture that valued small talk over substance. Michael was the envy of everyone in the room, and even though they understood all the money in the world could not buy them his lineup of Oscars, they never stopped wanting them.

I sighed to myself. Michael was the communicative and open father I wished I had.

Drinking wine and slurping oysters, I relaxed into conversation with a circle of guests that included the artist who designed the label. She explained that the art was a portrait of her wife. The *Post* editor laughed innocently and said she thought the image was intended to be Marilyn Monroe. The supermodel and the anchorman nodded in agreement.

"Something about the painting reminds me of a drag queen," I said, gulping down a glass of First Crush. "Either that or a rodeo clown, it's hard to say."

It took me a moment to realize that my comment caught the artist off guard. Perhaps no one had ever volunteered anything like that about her art. To cut the tension, I bet a hundred bucks that I was the only person there who had ever been to a rodeo. The shock in their faces changed to amusement and we all clinked glasses.

"How'd you get to be a rodeo expert?" the *Post* editor asked. "Off the record, of course."

The partygoers leaned in to hear over the din. "I grew up on a cattle ranch in South Dakota and my father was a rodeo star." But it was much more than that. I came from a conservative, religious household in a region of America where everything was different from New York: culture, education, food, income, language, work—the stuff of life itself. At its peak, my family owned almost ten thousand acres of ranch land. It sounded big-time in New York, where crops like the grapes I now grew were prized in small bunches. But for a cattle rancher like my father, it was a tough living. Born a poor rural kid, I left home at eighteen on a mission to forge my own identity. Now both sides existed in me: the wide-open Great Plains that had nurtured me as a boy, and the glossy city and coastal suburbs to which I had escaped as an adult.

"I had a pet horse when I was little and it bit a chunk of hair out of my head," I said.

Laughter turned to disbelief when people realized I was not kidding. That is how I had come to resolve the tensions of my childhood, by straddling the supposed line between city and country. The anchorman asked why, exactly, my horse had an appetite for hair.

"He got spooked by a rattlesnake and was mad at me," I said.

Everyone chuckled.

"It sounds like you lived *The Grapes of Wrath*, out there in flyover country," a bejeweled publicist said.

My childhood was amusing to East Coast types, and I dismissed their polite mocking as a form of fascination or curiosity. It was more interesting for them than talking about the stock market, which could only be up or down, never horse or rattlesnake. I nodded like I was in on their jokes, but this was the millionth time I had heard these lines. Flyover country was seen by many as a cultural wasteland, a ghetto so vast it was best to fly over it rather than drive through it and try to understand its people or natural wonders. I had shrugged off the term as a lazy way to think. Farmers and cowboys did not exist in the minds of most New Yorkers, except as stereotypes from a century ago. In fact, we did exist in modern times, just in places they never visited.

I once took a group of friends from New York City back to South Dakota and was stunned to see it through their eyes. The most frequent word spoken on our trip was *bleak*. Everything was worth commenting about—the cows, the horses, the rodeo cowboys, the Badlands, the halting speech patterns of ranch folk, the dried beef sticks in jars next to cash registers, the slot machines in gas stations, the absence of avocado toast from restaurant menus, and the way every adult drank watery light beer from cans.

Michael approached with his friends, one a top executive at HBO and the other a billionaire real estate developer who owned some of the more notable skyscrapers in the city.

"Trent, help us settle a debate. You're the tiebreaking vote. Which *Lord of the Rings* did you like best? The third one, right? It won the most Oscars."

"Nope. The second one."

"Wrong answer, I just lost my bet. Why?"

"I liked those giant talking trees that protected the forests."

Michael nodded. "Of course you did. The Ents! Spoken like a true botanist."

I took a car service back to Long Island. During the ride I couldn't shake the rough cadence of my father's voice on the phone. I had mixed feelings about his invitation to South Dakota—feelings I could not parse clearly. I arrived home well after midnight, making lots of noise coming in the door. When the dog sitter saw me, judging by the look on his face, I must have been a drunk mess. For part of the night I lay in bed awake wearing my suit as the room spun. Feeling sick, I stumbled to the beach as the sun peeked over Robins Island and cast an orange glow over the bay. My shoulders relaxed, my breathing slowed, and the buzz of the partygoers' laughter in my head faded away until the only sound was the gentle lapping of water onshore.

The view triggered an old, deep memory: the unobstructed freedom of a flat, wide horizon. It was not too late if I left now. There was still time. The prairie was calling, and I had to go.

CHAPTER 3

THE LONG WAY HOME

My puppy's jowls gave him the appearance of a young Winston Churchill. He watched me pack an overnight bag with his white face and floppy brown ears, though it was hard to tell which direction his lazy eye was looking. The first time I brought him home, he frolicked around the beach chasing seagulls with such a carefree spirit that I predicted his life would be full of adventure. So, I named him Caper.

He was the counterpoint to my experience with dogs as a young boy. My father thought dogs belonged outdoors with the livestock, curled up in hay bales for warmth. Many of our dogs died violent deaths on the ranch—run over by tractors, mauled by coyotes, kicked in the head by cows. The first time it happened, I cried, but I grew numb with every replacement puppy that my father brought home. I named all of them Walter.

Caper was no Walter. He hopped in the back seat of my car, which I lined with a quilt Mom sewed from my high school marching band uniforms. I had just picked up my silver BMW with fifty miles on the odometer, the first time in my life that I ordered a car trimmed with options and waited months for it to arrive from the factory. It was an upgrade from the Honda I had driven since grad school, with three hundred thousand miles and a dented hood, which was only reliable in the sense that it reliably skidded off the road after the faintest dusting of snow.

I planned to keep my visit short and sweet by staying two nights with my parents, then driving back to New York for winery meetings next week. There was a nostalgia about childhood that I had observed in my friends, the comfort they felt going home for the holidays to sleep in their childhood bed under the same roof as their parents. I couldn't relate to that feeling.

Caper and I had to cover about six hundred fifty miles a day to make it to South Dakota in time for Thanksgiving. The first day, I fielded press calls about the First Crush launch as the traffic snarls of New York and New Jersey gave way to longer and emptier stretches of highway. I drove through Pennsylvania and Ohio, stopping to walk Caper every few hours and admire the Appalachian forests, dense with maples and oaks. Hundreds of miles passed, always a straight track west on I-80. No sharp line divides the East Coast and Midwest, but the fallow cornfields out my window on the second day welcomed me into the heartland. Frozen duck hunting sloughs flanked the road throughout Indiana and Illinois, with countless farmhouses that were clean and white and fresh. When fatigue set in and I started nodding off at the wheel, I drove long stretches with the windows down. The frigid air kept me awake. It became increasingly obvious with every passing mile that I had not packed enough warm clothes.

On the third day, I crossed Iowa, where I-80 cut a double trench through winter's white all the way to the horizon. Staring down that monotonous road, I tried to imagine what it would be like to see my father again.

ON TOP OF being a cattle rancher and craftsman, he was a champion rodeo athlete and Vietnam vet. He was of average height but had the shoulders of a lineman and a barrel chest that stretched his Western shirts. He wore sideburns on his wide jaw, square-framed glasses, and a trimmed ring of brown hair around his otherwise bald head. Like most of the men I grew up around, my father was a little rough but cleaned up for Sunday church by shaving, slapping cologne on his neck, and tying a Windsor knot in his best necktie. Any other day, he would return from work at night with oil handprints on his coveralls, black smudges on his face, and sawdust clinging to his arms.

His hands were angular and nimble, with bruised fingernails and swollen knuckles, and a stumpy left thumb from a drill press accident. I had admired his hands when he worked in his shop, sawing wood and hammering steel shoes onto horses' hooves. I would cringe whenever he asked me to give him a specific tool for the task he was doing. His tools were foreign objects to me, and I never could remember if a Phillips head screwdriver was the flat one or not.

"I told you, it has a cross on top," he would say, articulating each word slowly in exasperation.

When I was a kid, I loved tagging along with my father while he did ranch chores. I rode in his lap while he cut alfalfa in the hard sun or fed cattle winter hay while stinging sleet pelted the ground—always with one arm around my waist, and the other hand clamped on the tractor's steering wheel. On days we drove around country roads with the windows down in his gold 1976 Chevy truck—which he nicknamed Old Yeller—he let me shoot beer cans off fence posts with his .30–06 Winchester.

He had a degree in animal science from South Dakota State University but was a cowboy by the power of his own say-so, which was more than enough. He broke wild colts by riding and whipping them until they stopped bucking and submitted to his commands. He was a man of few words, but when he spoke, it landed with authority. He was obsessed with the size, scale, and dimensions of everything, and spoke a language of measurements that awed and intimidated me. There were bushels of wheat, kernels per ear, miles per gallon, acres of alfalfa, widths of barn doors, points on a mule deer rack, feet of snow, inches of rain, calibers of rifles, gauges of steel, diameters of buckshot, and yards he crawled on his belly through sagebrush to shoot that seventeen-inch pronghorn buck. I feared I was not man enough and did not measure up in his eyes, both in terms of my determination and my physical stature as a young man.

Before my family's cattle ranch went belly-up during the farm crisis of the 1980s, we fit the stereotype of both a people and a place: ranchers on desolate, flat earth. Our corner of South Dakota covered an area roughly the size of Connecticut but was inhabited by about three thousand people. "The middle of nowhere" was not a cliché, it was a fact. There was a published study that made the rounds, written by a statistician at the University of California, Berkeley, declaring the latitude and longitude of the most isolated place in the continental United States. Those coordinates pointed to the land just south of my family's ranch. The author's rationale was that we lived one hundred forty-five miles from a McDonald's drive-through, farther than any other Americans.

Most residents of the area were scattered among ranches and tiny settlements with simple names like Bison, Meadow, and Wall. The town I identified with most was the one that had a movie theater on Main Street, a lumberyard where my father bought wood, and a livestock auction where he sold cattle. That town was called Faith. Our ranch was north of Faith, thirty miles from the post office, forty miles from church, and sixty miles from the hospital where I was born.

I grew up with a nagging feeling that life was certainly elsewhere. I yearned for movement and escape. My father and I occupied our days by riding horses or hiking across the prairie, always with our heads down scanning for rattlesnakes. I was also constantly on the lookout for some artifact or remnant to prove that a nomadic tribe or ancient glacier had been there before us. When I did find something, like a seashell fossil or arrowhead, it made me believe that my family was not alone, and that something important had happened on this land. A few things had. Our ranch was near the site where a grizzly bear mauled Hugh Glass in 1823, an incident that later inspired a movie called *The Revenant*, starring Leonardo DiCaprio. We also lived a few miles from the spot where Robert Pirsig camped out on his road trip in the book *Zen and the Art of Motorcycle Maintenance.* I asked my father once if he met Pirsig when he passed through the area, and he mumbled something about "them danged hippies."

It was not all bad, being so isolated. Though money was tight, my parents knew how to build and grow things, so we always had our basic needs met. The ranch was a master class in food production: we had a rambling vegetable garden that stretched for an acre beside the barn, the grain-fattened steers we butchered, egg-laying hens, and thousands of acres of durum wheat. My father grew alfalfa that he cut and baled to feed a thousand head of cattle over winter. Mom ran the house, cooking for our work crew during the harvest season. She would drive tubs of fried chicken and thermoses of iced tea to the dusty fields and wait at the end of the windrow for my father and the grain combine drivers to take a break. She kept the root cellar stocked with vegetables she had plucked from the garden and canned, and two deep freezers packed with beef and venison wrapped in crinkly butcher paper. Mom did not talk about the isolation of ranch life, but I observed it whenever she baked desserts. She would stockpile the ingredients so she would not have to drive forty miles to the store for flour when she wanted to make cookies, and she often baked extra to bring to church on Sunday.

The flat earth was more of a full-immersion experience than a view. I had dirt on me every day. My feet would get stuck in the manure-caked corrals so that I would have to slide out of my cowboy boots and trudge home in my socks. Gravel lodged in the skin of my elbows when my horse bucked me off. After a bath, I still had dirt in the folds of my ears. I pulled radishes and strawberries out of the garden, rubbed them on my jeans, and ate them on the spot. Mom said a little dirt was good for me.

My sister, Lucinda, was two years older than me. In the summer months when we were kids, we staged daily foot races between the swing set and the one-room schoolhouse. I was always the slower and shorter one, until one day I gained a slight advantage. Lucinda tripped and fell halfway through the race and I sprinted ahead with my arms raised in triumph. Little did I know that my victory would be so hollow. Mom said Lucinda had a seizure, though I did not understand at the time what that meant. From that day forward, Lucinda stumbled almost every time we raced. Eventually, she stopped running altogether, and walking beside me became a monumental struggle. Lucy, as I called her, was slow to develop some sounds, like the letter *R*. She stuttered, so I learned to translate for her. Isolated together, we developed a way of communicating that went beyond speaking.

For all my fears of not measuring up to my father's outsized masculinity, I had the opposite fear about Lucy. I could not understand why I grew taller than her even though she was older, why she fell behind in school and I excelled, or why I ran free while she sat in a wheelchair. My father would try to explain the sober medical answers to those questions, but they were of no comfort. Providing comfort was not in his nature.

Since I had left the prairie, my father and I had become more and more *estranged*—a word that meant different things to different people. For us, it was not anything we formalized or kept score about. I just didn't feel like seeing him or hearing his voice for long stretches of time, so I didn't—not that we had much in common to discuss anyway. I knew very little about my father or his own boyhood. He never asked me questions about my life and he never talked about himself, either. All I had to go on were hints from Mom and my own observations, studying every little thing he said and did for clues—which wasn't much. Whenever I ached to know more, to have a real conversation with him, bitterness consumed me. I distracted myself with work until the feelings passed. Stoicism was a survival skill that I picked up from him early on.

After I had moved as far away as I could for grad school in New York, the little silences between my father and I happened more often over time. They added up to something bigger, something I called estrangement. It was easier to live an independent life two thousand miles away from a past I had hoped to forget. What did I have to show for all of it? My attitude had softened since my divorce. I had to admit that I had lost time with my father and probably hurt Mom even more. This road trip felt like a new beginning, a chance to hit the reset button. I wondered if my father felt the same way.

AT THE END of the third day driving, I reached the South Dakota border. Darkness fell early in the afternoon this close to the solstice. I was so starved for real food and so tired that I kept swerving onto the snow-packed rumble strips on the side of the road. I heeded my father's advice from my first driving lesson as a teenager: don't overbrake, don't oversteer, and coax the car like a kite until it straightens.

I stopped in a border town with more gas pumps than people. Eighteen wheelers idled in the parking lot of a greasy spoon diner. Haggard truck drivers looked at me suspiciously while I devoured a burger and typed emails on my laptop. Caper and I slept a few hours in a thirty-nine-dollar motel room and left the truck stop before dawn.

The sun rose in my rearview mirror to mark the fourth and final day. Morning cast pink cotton candy streaks across the sky, wider and bigger than any sky back east. Farmhouses were fewer and farther between now, often small specks in the distance. I exited the highway and drove on back roads with no sense of direction other than west, snacking on beef jerky from the truck stop. Movement in the distance caught my eye. It was a herd of pronghorns crossing a wide, flat basin of rangeland. They were North America's most elusive animal, built for endurance and capable of outrunning any predator. I pulled over and got out. Caper tugged at his leash. The pronghorns streaked across the road and disappeared into the great, empty remoteness of the prairie. The size of the sky settled on me. There was a frozen silence in the air caused by sheer magnitude that made my ears ring.

Back on the road, long bands of snow on the prairie resembled deep swells on the Atlantic—a similarity that only revealed what little else these two places had in common. The landscape was dotted with black Angus cattle and the ditches were framed by infinite rows of barbed wire fencing strung between crooked wood posts. What few trees existed here were not native but had been planted by homesteaders as windbreaks to mark the flat space between horizons. The trees grew straight out of the prairie sod with no lush undergrowth like in the eastern forests. It was hard to see autumn with no leaves.

Every familiar dimple in the landscape and every remembered barn just off the highway twisted my gut. Coming home—the math stunned me, how many years I had been away. So much had gone untouched: the ranches, the deserted towns with one stop sign and a church, the desperate public service billboards that read FOR GOD SO LOVED THE WORLD, and EAT STEAK, WEAR FURS, KEEP YOUR GUNS, THE AMERICAN WAY.

Roadside attractions were few, but each one distracted me enough to calm me in the final hours before seeing my parents. There was The World's Only Corn Palace—an entire basketball arena constructed out of harvested grains—and the world's largest pheasant: forty feet high, twenty-two tons of painted concrete. There was the historic marker where Meriwether Lewis and William Clark met Sacagawea on their famous expedition in 1805. There was the decommissioned Minuteman nuclear missile silo and the moonscape of Badlands National Park. At the last of these quirky midcentury landmarks, I pulled over for a walk near a familiar billboard that read WELCOME TO THE WORLD'S LARGEST PETRIFIED WOOD PARK. As a kid, I had spent many summer days playing hide-and-seek among the Cretaceous tree trunks. Caper weaved his way through the park with his nose to the ground.

This entire region was submerged by the Western Interior Seaway seventy million years ago. When the waters dried up, they left behind a treasure trove of petrified wood and aquatic fossils—tokens from an alien world. In the time I had been gone, though, someone had erected a new feature here: a small-scale model of Noah's Ark, sitting at chest height, with little wooden animals waiting to board. Elephants, tigers, oxen, sheep, pigs—all the way down to a tiny pair of chickens—stood patiently in line. I slipped a piece of petrified wood into my coat pocket and rubbed its smooth, cold surface while reading the ark description on the informational placard:

> *Noah built the largest wood structure in the world using standing dead timber. God established a covenant that if Noah built his boat, the Earth would never flood again. The waters took a while to recede, and Noah sent out birds to see if there was any dry land. A year later, the birds flew back to the boat with word: The Earth was ready for him again. After the flood, Noah planted a vineyard, and was the first person to discover wine.*

It was a reminder of everything I had learned growing up in my family's Young Earth church. At summer Bible camp, the minister explained why we had abundant fossils in the area: the dinosaurs were too big to fit on Noah's Ark, so they drowned. It seemed like a logical explanation at the time, but I asked the minister why Noah left the trees to die and how he fed all those animals on the ark without any plants. He shrugged and recited Psalm 96:12. "Let the field be joyful and all that is in it. Then all the trees of the woods will rejoice before the Lord."

I viewed the world through trees—not just the fruit, nuts, wood, shade, and autumn foliage, but the whole trees themselves, the organisms, the living beings. My earliest memory of my father was from the day we planted a ponderosa pine in the barnyard using a free sapling from the Arbor Day Foundation. It arrived cradled in a delicate tar paper pot with a foot-high trunk, flexible and green, the diameter of my father's pinky finger. I held it in my small hands and felt the mass of eager roots that had punched through the bottom of the pot, searching for water, and freedom. Its only whorl of needles, dark green and long, oozed sticky sap. My father dug a hole under the summer sun, sweating from the exertion, his jaw clenched. When it was wide and deep enough, he knelt at the edge of the hole and told me to set the little tree into it. He pushed the dirt back in with his spade while I chopped at clods with a small trowel, and then we tamped down the earth using the heels of our cowboy boots. My father administered the sacrament of first water, poured from a feed bucket he had dunked into the cattle tank. I stepped away from the planting, underneath his waiting arm.

"Wait and see, Trent, someday this'll be bigger than you," he said.

Caper and I left the Noah's Ark exhibit. Before driving the rest of the way home, I texted Mom.

"Almost there! Just want to stop at the ranch first."

My parents no longer owned our ancestral ranch. After the farm crisis, we sold it and moved to town. I had not seen it since I left for college and was curious how it looked after all these years.

Mom texted back right away: "Dad says don't stop at the ranch."

"Why not?"

"You'll be late for supper."

I did not want to argue. I could swing by the ranch on my way back to New York in a couple days.

"Alright, I'll come straight home."

Over the last rise to the west before the turn off to my parents' house, my pulse shot up. I squinted for the horizon's lone indicator of the road that crossed the railroad tracks, the small sign that bore its appropriately literal name: West City Limits Road. It was the road to a little beige house on the west side of a small town in a large state on the edge of winter. It was a place where dirt turned into pavement, where the wild prairie tapered and gave way to corner stores and cul-de-sacs, a place I used to call home.

CHAPTER 4

THANKSGIVING

It was three o'clock but would be dark soon enough. My breath tasted like beef jerky, my heart was racing, and my throat was tight. Caper and I walked up the driveway to my parents' house and, as if by magic, the garage door jolted open and creaked up its track. I was surprised that it made such a racket. My father used to keep that chain so well lubricated that oil dripped on the cars. In the garage, his truck was surrounded by the dusty cardboard boxes and rusted cans that he always hated to throw out. They ranged from Folgers to Valvoline to Quaker Oats, saved for decades—who knew when you might need one just the right size? His handwriting was scrawled on some of them: truck parts, rodeo gear, mismatched nuts and bolts. He seemed to be midway through a reorganization project.

Next to the spot where my father's Winchester rifle with hardwood grips leaned against the wall, Mom had stacked Tupperware waist-high, each containing a different flavor and shape of Christmas cookie. My mouth watered at the sight. As always, she baked enough to feed an army and put them in the garage to freeze.

Here goes nothing. I stepped into the house. My father was asleep in his recliner with an NFL game on the TV and Mom was cooking in the kitchen.

"Welcome home!" Mom said.

Caper immediately made himself at home. He darted between my legs and ran around the house, knocking a glass ornament off the Christmas tree. Mom had decorated it earlier than she used to when I was a kid.

"Sorry about that, he's a little hyper," I said, rushing over to pick up the broken pieces of glass on the floor. Mom had always kept the house tidy herself; hiring a cleaning lady would have been an extravagant luxury. She had a knack for making it appear that we had more than we did by decorating the house with things that were handmade and well arranged. The tchotchkes, heirloom glassware, hand-sewn quilts, and honey-stained oak furniture I had grown up with were all there but reshuffled. Everything looked familiar but off.

My father awoke with a start when Caper jumped into his lap.

"Hello there, puppy," he said, sounding happy despite his hoarse voice.

He did not look at me or say hello. He was too busy scratching Caper's ears while Caper licked his face. I was surprised and a little resentful at their instant affection. Caper accomplished in minutes what I could not in years.

I threw the glass shards in the trash and hugged Mom. She had always been petite and still was. She liked to remind me that she weighed ninety-six pounds on their wedding day in 1968. Her hair was permed, coiffed neatly, and thinning with age.

"How was the drive?" she asked.

"Alright." I shrugged. "I saw that crazy replica of Noah's Ark. No wonder they named the town Faith."

On the way to the bathroom to wash up, I passed my framed eight-by-ten high school graduation photo hanging in the hallway. I wore my father's too-big navy blazer with awkward shoulder pads. My skin had been airbrushed clean of acne and the whites of my eyes were doctored to look brighter, but the sizable gap between my front teeth was untouched. The faint black dot on my left earlobe provided the only evidence that I wore an earring when my father was not around. I had wanted a piercing in high school, but the straight football quarterback I was hooking up with on the down-low told me that piercing my right ear would mean I was a faggot. So, the Claire's boutique lady at the mall pierced my left ear instead, a sure sign I was absolutely, positively *not gay*. I had run so far and so fast away from that boy in the photo that I barely recognized him.

"Let's eat," Mom said.

My father stood up from his recliner with a groan meant for no one and shuffled to the table. The years had not been kind to him. He had lost a lot of weight and was hunched over. The pilled navy sweatpants and maroon sweater he was wearing were wrinkled like he had slept in them, and they hung off his body like he was a scarecrow. One thing had not changed: we sat down to eat without exchanging pleasantries. He gave me a subtle nod and a subdued hello through pursed lips.

After my father said the prayer to bless the meal—*Come, Lord Jesus, be our guest*—I uncorked a bottle of merlot from my vineyard. The first time I ever tried wine was also on Thanksgiving, thirty years ago, when my father poured me a shot glass of Cold Duck, a cheap sparkling wine that looked expensive with its purple foil wrapper. I used to drink it so fast that fizz came out my nose.

I poured glasses of the merlot for all three of us.

"This was served at President Obama's inauguration," I said.

"I ain't drinkin' no Obama wine," my father said, pushing away the glass.

"Thanksgiving is bipartisan, Dad. Cheers."

I lifted the glass to my nose and inhaled the aromas of stewed blackberries, licorice, and tobacco. The wine had improved with age: its flavors were more expressive, and the bitter tannins had softened slowly over time. This was the last bottle remaining of Bedell Cellars 2009 Merlot, the official red wine of the 2013 Inaugural Luncheon, which was held under the Capitol Rotunda minutes after the swearing-in ceremony. Members of Congress, Supreme Court justices, and former presidents and first ladies in attendance drank this very same wine.

"And they paired it with a filet of South Dakota bison for the main course," I said. It was one of the highlights of my career, something I was proud of. I leaned forward, searching my father's face for some crease of shared conviction that this wine—*my* wine—was something he could be proud of, too. He peered back at me with his hazel eyes and square gunmetal eyeglasses that covered half his face—the same frames he had worn since the 1980s. They resembled the wire-framed glasses I was wearing except I had paid a ridiculous six hundred dollars for mine at a boutique in SoHo. He shook a few pills into the palm of his hand and downed them with a gulp of merlot.

"Purdy dang good," he said.

A few minutes passed with only the sound of silverware scraping on plates.

Mom's beef stroganoff was comfort food. One summer day when I was eleven, I sat at the dining table devouring stroganoff while Mom picked cactus needles and cockleburs off me with tweezers. I had skinned elbows and blood spattered across my T-shirt. Outside, my horse, a stubborn chestnut named Socks, was tied to a post and lathered in sweat. He had bucked me off when I was out riding in the pasture with my father.

Mom's voice broke my spell. "I didn't have time to make a turkey this year," she said with a sigh of resignation. "It's just too much."

Caper sat at attention next to my father, chewing a piece of meat.

"Don't give him people food. He's not some *farm* dog," I said.

"Relax, dogs like treats and he's too skinny."

"Yes, but he's *my* dog."

My father cleared his throat and put down his fork.

"So, whatever happened to that boyfriend of yours?" he asked, smoothly, without missing a beat.

I was shocked. In my thirty-seven years, my father had never asked me about a relationship of any kind, with men or women, romantic or platonic.

"You must mean my ex-husband, since you didn't come to our wedding," I said.

"Yup, that guy. Mom told me what happened."

She interjected: "Is that what you even call it, though? A *wedding*?"

I started to get angry with her for what seemed like a homophobic comment, but it was a fair question. After dating for three years, my boyfriend and I held a ceremony for a hundred guests on a beach in Spain, which recognized same-sex marriage. The problem was that I wasn't a Spanish citizen, my boyfriend wasn't an American citizen, and same-sex marriages performed outside the United States were not recognized everywhere back home. Our union was still in legal limbo at the time we separated, so it was easier just telling people I was divorced.

"Would've been nice to hear from you, at least," I said to my father.

"I was getting chemo at the Mayo Clinic last summer. Doc said I couldn't fly."

Before the phone call that precipitated this visit, the last time my father had called was seven years ago, to tell me he had been diagnosed with colon cancer. In the years since then, we didn't have another conversation about it. Mom told me he underwent surgeries, radiation, and chemo to shrink the tumors. I hadn't heard anything otherwise and assumed no news was good news.

"What happened with your ex?" he said.

"I'll spare you the details. I had to leave him. It was for my own sanity. I was scared. I lived out of a hotel room for a while before adopting Caper and moving into the beach house."

"Would've been nice to know."

"Since when are you interested in my love life?"

"If my son is hurtin' I oughta know."

I was surprised by the entire conversation. It did not jibe with the cold and distant person I expected him to be. I choked down a bite of food and muttered, "Damned if I do, damned if I don't."

"You ain't got time for jokers," he said.

Mom and I cleared the plates without speaking and my father went back to his recliner. Caper jumped in his lap. The house was completely silent, country silent, and the sun cast diagonal venetian blind stripes across the furniture and floor. I sat on the wine-colored sofa opposite my father and opened my laptop. He opened a Bible in his lap. He had a presence about him, something solemn but powerful. His hands were thick and leathery, the hands of a man who had been hard at work all his life, and they grasped his Bible firmly. I polished off the bottle of merlot and read the *New York Times* while he read Scripture.

The Book of Job was my father's favorite. He upheld Job as the prime example of having godly patience while suffering. Job's life was supposed to teach us about our own lives in three ways: humility (God was in control); steadfastness (Job refused to curse God despite his awful circumstances); and reward (Job's life turned out better in the end, after all the suffering). Although I was not religious as an adult, I retained a surprising amount of Scripture that I had been forced to memorize in catechism as a teenager. There was a section of Job, chapter 39, that stuck with me over the years: "Have you given the horse strength? Have you clothed his neck with thunder? . . . He mocks at fear, and is not frightened . . . He devours the distance with fierceness." I told my father once that if he had a spirit animal, it would be a horse. He dismissed my characterization as woo-woo hippie bullshit.

My father closed his Bible and quoted Job 1:21. "The Lord gave, and the Lord has taken away, Trent," he said. "Blessed be the name of the Lord."

He turned on the TV and we watched the game for a few minutes, the Philadelphia Eagles versus the Dallas Cowboys.

"Are you okay?" I asked him, raising my voice to be heard over the TV.

He didn't hear me. He was already asleep.

There had been a time in life after we sold the ranch when my father could not sleep at three in the morning and he would go work in his shop, reloading ammo or fixing a broken drawer on a cabinet. Mom would say he was putzing around. Sometimes I would wake up and go watch him in my pajamas until he left for his welding job at five o'clock. All that restlessness, all my father's hard living, must have been connected to the exhaustion that now left him sleeping in his recliner in the late afternoon sun. I didn't follow football, but I sat there and watched TV while my father and Caper snored. There was something comforting about it.

WHEN I WAS a kid, we had one of those TVs in a giant wooden console with fabric speakers and a big knob to turn channels. We got PBS somewhat reliably, and CBS if the weather was clear, but nothing else. The fuzzy pictures came into focus when my father climbed onto the roof and aimed the antennae toward the cities in the east. This was the only way I got sketchy messages that the outside world existed, in my childhood before cable TV and the internet.

More than any other show, we would watch a country music sketch comedy called *Hee Haw*. The host, Minnie Pearl, would stand onstage telling the audience how proud she was to be there wearing her best dress, but she had forgotten to take the $1.98 price tag off her straw hat. She would say that the price tag was symbolic of all human frailty, and my father would laugh so hard that his eyes squinted shut and filled with tears, and his bald head turned beet red. I had not heard that laugh in so many years.

We also tuned in every week to Mutual of Omaha's *Wild Kingdom*, hosted by Marlin Perkins. I was fascinated by the strange animal behaviors that show documented, like fish that changed sex when they needed to, or spiders that ate their partners after mating. My father had no qualms about me watching a lion suffocate a gazelle in its jaws, but as soon as the elephants started having sex, he would tell me to go to my room. I would sit on the edge of my bed and through the thin walls of our house I heard the host whispering off camera, "Here we see the male elephant mating with the female and ensuring the continuation of his lineage." Once the bull elephant dismounted, my father would call me out of my room. After the show, he climbed back on the roof to remove the antennae, lest it become a lightning rod during a thunderstorm.

I WAS BEAT from all the driving and it was already pitch-black outside. I could catch up more with my parents in the morning. I carried my overnight bag to my old bedroom and discovered that they had transformed it into an office. Bookshelves and a desk with a computer now sat where my bed used to be.

"You'll have to sleep in Lucinda's old room," Mom called from the kitchen.

They had left my sister's room intact all these years, with the same pink floral quilt and pillows. Lucy had the hardest time sleeping. Sometimes I would wake up hearing her scream my name through the wall that separated our bedrooms.

TENT! she would cry, without the *R*.

I would jump out of bed and run to her room, finding her tossing and turning, tears and snot on her face, sobbing. Her sandy blond hair was wet against her hot forehead and I would pin it back with her favorite purple barrette, trying to comfort her, but she would look past me and scream and make a gargling noise, as if she were choking. My father would rush in and carry her to the bathroom with a gaunt, defeated look on his face. I would stand in the doorway watching Mom tend to Lucy, both of my parents' faces hollow, their eyes swollen with worry and exhaustion. My father would close the door and I would listen until Lucy's gasping stopped and he carried her back to bed. I could not say how long it lasted or how often I had seen the same episode repeated.

I slipped on sweatpants and got in Lucy's bed. My feet hung off the end of the mattress. Muted notes of a familiar church hymn drifted up from the basement: the Common Doxology. *Praise Father*, *Son*, *and Holy Ghost*. Mom was practicing the organ. Sitting on the doily on the nightstand was a little framed photo of Lucy in the wheelchair races at the Special Olympics. She looked triumphant, smiling a smile fit to sell toothpaste, with a medal on a ribbon hanging from her neck.

There were other kids around her in wheelchairs on a high school track with painted red running lanes. The photo leaned against a glass vase containing a long-stemmed white rose: dry, brittle, and dusty. Next to that, there was a heart-shaped needlepoint Mom had sewn with pink thread on white cotton. The stitching read: *A mother holds her child's hands for a moment, but their heart forever.*

When my eyes would not stay open any longer, I switched off the dim, low-watt bulb next to the headboard. Outside, all that was left of the prairie was a black sky and thin crescent moon. I smiled.

"Good night, Lucy."

CHAPTER 5

THE LAST COWBOY

Something awakened me, though inside there was only the sound of the furnace being startled into service. Outside it was frozen and quiet, but for the wind. The clock said 7:30, give or take. I was not going back to sleep. A sepia-tone painting of Jesus Christ hung on the wall, staring at me, his brow drawn and imploring. Where was I? I was not in my beach house; there were no seagulls, no crashing surf. This was not New York City, not Long Island. The shadows of my sister's old bedroom came into focus. Oh, I was in South Dakota, the only place in America with five times more cows than people.

I needed some fresh air to clear the melancholy out of my head. I pulled on thermal gym clothes and tiptoed out the patio door with Caper to avoid waking my parents. It seemed odd that the lights and TV were on in the living room, but the volume was muted. Perhaps my father had gone to bed late and forgotten. I turned everything off.

Bracing myself against the cold oncoming wind, I ran through knee-high tumbleweeds that piled up beside the railroad tracks. I jogged between the rails, stepping on wood ties to pace out a rhythm.

Fog wrapped the tops of wayward juniper trees in the creek basin on the other side.

ONE HUNDRED EIGHT years ago, in 1906, my great-grandparents on my father's side rode a train down these tracks on their journey to the western frontier. They took advantage of President Lincoln's Homestead Act, which promised free land to anyone brave enough to occupy the immense area west of the Mississippi River. That land had been inhabited for millennia by Native peoples, of course, but those tribes and the roaming herds of bison they depended on had been largely annihilated. The United States government had given massive swaths of land to the railroad moguls to enable poor white settlers to go west and create commerce. My great-grandparents, Jake and Christine Preszler, could have stayed in New York City after crossing the Atlantic from Odessa, Ukraine. Instead, they ventured west to the last stop on the railroad at a frontier settlement called Faith. According to family legend, they got off the train and immediately felt at ease: Faith reminded them of their own childhood homes on the Black Sea steppe.

They raised eleven children in a sod house without electricity or running water. Grandma Christine lived to see her hundredth birthday. I didn't know many other details about my family history. When I was growing up, my elders never talked about themselves or their past. I absorbed the awareness that our family had grown and built things on the earth for a long time, and the railroad tracks I was now running along were their only link to the world beyond the isolated prairie.

The freight trains that moved through South Dakota were unable to stop in time if something got stuck on the tracks. A hundred cars of wheat could stretch more than a mile and weigh twenty thousand tons. I experienced that the hard way while in high school. I was riding with the church minister and another kid heading to summer Bible camp when the car's engine stalled. The minister jammed the transmission trying to force the car into gear while it was moving, and we rolled helplessly to a stop in the middle of the tracks. As he tried in vain to restart the car, we heard the airhorn blast of an oncoming freighter and everyone scrambled out, trying not to panic. Together, we pushed the car across the tracks, and, less than a minute later, the train barreled through. We were shaken up in the moment but laughed about it later with ice cream cones in hand. When my father heard about the ordeal, the important takeaway was this: I was not meant to die that day.

"The Lord will call you home when it's your time, and He wasn't ready for you yet," he said.

Now, in the distance, I spotted a horse standing in a treeless pasture behind a barbed wire fence. He resembled my father's old horse, Chili, with a buckskin coat and a white blaze down his nose. Even from afar I could tell by his immaculately groomed mane that he had an attentive owner. He watched me as I jogged past. After I turned to go back, a low, rhythmic pounding grew louder behind me. The horse galloped down the fence line with his blond tail flowing in the wind. He caught up and trotted alongside me and Caper for a few minutes. The horse looked like a powerful avatar from another era, and in his presence I ran faster. When he reached the corner of the fence, he stopped and snorted, pawing at the ground. I slowed down and looked back at him. It had been many years since I had touched a horse, so I approached him warily, certain that as I moved closer, he would get spooked. Instead, the horse sniffed my shirt, leaving a long, wet stain. I yanked up a clump of dried grass and held it out. The whiskers on his chin tickled my palm as he ate it.

Back at the house, it was eerily quiet. I walked into the kitchen where I found a handwritten note:

Took Dad to the hospital, come when you get this. ~Mom

My heart dropped. In a matter of minutes, I had stuffed everything back into my overnight bag and got in the car with Caper. I was in a panic, wondering what had gone wrong. I called Mom's cell phone a dozen times, but she did not pick up. I drove an hour to the hospital and parked in the visitor's lot, leaving Caper in the car with a bowl of water and the windows cracked. I arrived at the hospital information desk breathless. The receptionist directed me to the third floor—*Cancer Unit.*

All my life, a single image of my father had been seared into my brain. It was a grainy, sepia-tone photo of him from the professional rodeo circuit in 1968, his championship year. I kept the photo on my nightstand during the years we were estranged. In it, he wore a white felt cowboy hat and white button-down shirt with starched creases up the sleeves. Over the shirt he wore a khaki canvas vest with three buttons that were splayed apart by his barrel chest. Framing his left hand, clear as day, was a white-faced Timex watch with a black leather wristband. Behind him in the distance, bleachers rose into the sky holding thousands of rodeo spectators. Despite his crisp appearance, he was sitting on top of a several-hundred-pound black calf. He had pinned the animal to the dirt with his knees and clenched its hooved feet together with his meaty hands. The calf lay helplessly on its side with one eye in the dirt and the other eye peeled wide open in distress. My father was in the act of tying a rope around the calf's feet to immobilize it during the calf roping competition. I compared myself to this photo countless times throughout my life, and I was always the less masculine one, the less rugged. Our hands, though, were a different story, and the one feature we had in common. They were the same size and shape, and our fingers moved in similar restless mannerisms.

That long-standing image crumbled when the nurse led me into my father's hospital room. He stood at the sink in the bathroom wearing a thin cotton smock, robin's-egg blue and several sizes too big. The strings had come untied and dangled at his sides, so the back flapped wide open, exposing his sagging body, the small curves of flesh below his waist, and his bony, rail-thin legs. I covered my eyes out of embarrassment. Mom had been sitting on the sofa against the window and when I came in, she walked over to tie his robe shut.

"Morning," he said while slathering his face with white shaving cream. The tone of his voice had an upbeat, no-big-deal perkiness that contradicted the entire scene. I was totally confused.

"What's going on? Are you alright?"

My father shaved in slow strokes and rinsed the razor under hot water. It was not unusual for him to be shaving, even though he had been admitted to the hospital. I could imagine him stubbornly demanding a razor from the nurse, regardless of his medical condition. He had always taken meticulous care with his appearance, and I rarely saw him with a stray whisker or five-o'clock shadow. Whenever I noticed hipsters in Brooklyn with hair grown unkempt across their faces to project a rugged, outdoorsy style while working on their laptops in coffee shops, it triggered an image of my father, who shaved every morning before dawn to go weld metal and rope cattle.

"I woke up in the night sick and thought it was the merlot," he said, pausing to take a long swipe of the razor across his jaw. He rinsed it off. Steam rose from the sink. "Doc thinks the cancer is back."

"What are they going to do? How serious is it?"

"Don't know yet. Have to wait for test results." He finished shaving and patted his face dry with a towel. The nurse helped him climb into the hospital bed, wheeling his IV cart alongside. Once he got situated, he looked around the room. "Where's Caper?"

"In my car—probably freezing."

"You got a hay bale for him in that BMW?"

"There's a hotel across the street, you should get a room," Mom said. "This is gonna be a while."

At the hotel, the receptionist gave me a room on the edge of the parking lot so I could easily walk in and out with Caper. She was a parishioner at the church my parents attended, and her husband was my Bible study leader in high school. She handed back my driver's license and squinted at my face.

"I remember you, kid. There'll be no charge today."

Back in my father's hospital room, nurses came and went. They drew blood, wrote on clipboards, and whispered to doctors. My father thumbed through *National Geographic* magazine with IV tubes dangling from his arms. I fumbled through an attempt at small talk to distract from the reality of what we were all facing.

"So . . . what are you reading? Planning your next hunting trip?"

"Nah, I'm done killing things," he said.

This declaration was so unexpected that the man in the hospital bed could not possibly be my father. Hunting and fishing were just about the only things we had in common when I was growing up. What little quality time we spent together was most often outdoors in nature, riding horses, shooting guns, and casting rods. From the day I was old enough to hold a gun until I left for college, we never missed the opening day of duck hunting season—not to mention the season openers for beaver, deer, goose, grouse, pheasant, pronghorn, and walleye. If it had fins, fur, or feathers, chances are I had bagged a few with my father.

"But I saw one of your guns in the garage," I said.

He shook his head. "I don't even shoot the rabbits in your mom's garden anymore."

"Which I'm not happy about," she said, wagging her finger.

I held up my phone and showed him a photo of the horse I ran beside that morning. It was easier than trying to tell him that I loved him. He tilted his chin upward to see through his bifocals.

"I'll be galldang, it looks like Chili, but that old boy's been dead twenty years," he said.

We fell into a comfortable silence. After lunch and a break to walk Caper, I sat in a folding chair next to my father's bed and clattered away on my laptop. Mom stitched a quilt square. I was stressing out about work; it was Black Friday, the busiest shopping day of the year, and back at the winery, bottles of First Crush were flying off the shelf. My father saw me checking my watch.

"I know you have to get back to work, and you'd go stir-crazy if you stuck round here," he said. "Besides, I'll be fine."

"Yeah, um, I was going to stay one more night and hit the road early tomorrow." I was sheepish about sticking to my original travel plans.

My father's eyes closed. He dozed off, the magazine still upright in his hands. Twenty minutes later, he awoke with vacant eyes and an open mouth while he stared at the plastic ticking clock on the wall. Mom and I exchanged a concerned glance but did not say anything. The doctor walked in and asked for privacy so he could examine my father behind the blue curtain. He handed me prescriptions to pick up from the hospital pharmacy.

"Will they release these to me, since they're in his name?" I asked.

My father shouted from behind the curtain. "Just tell them who you are—*tell them you're my son*."

Hearing those words gave me goose bumps. He had disowned me fourteen years ago, so our differences should have been insurmountable. Maybe this was a little crack in the door. Picking up the medications made me feel useful, like I was doing something tangible to help a situation that was beyond my control. When I came back, the doctor was gone and Mom was standing beside the bed, holding my father's hands.

"What's the news?"

Neither one answered me. I read their silence as an indication that they had received bad news and were afraid to tell me the truth. In the same split second that I began to confront this reality, I pushed the facts of what I was witnessing out of my mind. I could not let myself believe my own eyes and ears. So I chose to let myself believe other things instead—such as *my father is a tough cowboy*.

I set the pill bottles on the bedside table and stood awkwardly, with my hands in my pockets. Televisions blared from other patients' rooms, tuned to different channels. There was a purple bruise on my father's arm where they had inserted another needle. My parents made eye contact with each other but did not say anything. Their silence lasted quite a while.

I studied my father's worn-down face and the tubes trailing out of his arms. The sight of him in a weakened state bled me of my bitterness. I understood then why I had not come home to visit in so many years. It was not just that he didn't approve of me as a gay man, although that was our biggest barrier. It was also that I had been afraid of how I would feel if I ever saw him sick or weak. He was always the stronger one, and if he was diminished, my weakness might be exposed. I didn't know how to play the role of the stronger one, and worse yet, I had started to fear that I was not a good son.

My father cleared his throat and broke the silence, finally. "Blood tests came back high. Liver can't handle more chemo."

"I'm sure they have other options," I said, slowly, confidently. "And besides, you're tough, you've been through worse, right? How long are they keeping you here, anyway?" Hearing my own voice, I realized that I had adopted the same optimistic tone that he used on me when I first walked in the room.

"That depends. Got more tests tomorrow."

It was getting dark outside. I would have to start driving at the crack of dawn and make fewer stops along the way if I were going to be at the winery Monday.

"Well, how about this," I said. "I'll come back for Christmas so we can catch up more when I'm not so jammed with work."

"That's fine," my father said, struggling to keep his eyes open.

"You should get a good night's sleep," I said, leaning over for a goodbye hug. I hadn't hugged him when I arrived, so I had to try now, even if he resisted.

My father seized me into his chest. He smelled sour in the way people on medication sometimes do. He buried his face in the crook of my neck, and he sobbed. His hands dug into my back like he was making fists.

I took a breath to say something, held it for a few seconds and exhaled, but no words came out. I wanted to say *I love you*, but I couldn't. I had never heard him say those three words my entire life. I wanted them to knit together in his mind and for him to deliver them to me too.

It was the only time he held me as an adult. Then again, in the seawall I had built between my present and my past, I had mostly forgotten how close we once were. Being his son, I had no choice but to be scared with him.

He did say three words, but they were not the ones I hoped for.

"Safe drive, okay?"

And then he let me go. I stood up.

"There's a box in the garage that I want you to take," he said, adjusting his bedsheet. "It's some stuff from the ranch. I was gonna save it for Christmas, but—" His lower lip quivered.

It was difficult to imagine a man like my father—proud and strong, impaired by cancer—sorting his things into boxes and cans in the garage. How would he adjust if the treatments dragged on and on? Could he live a happy life if he were unable to grasp a hammer or ride a horse or shoot a gun?

"Look, I'll be back in a couple weeks. I'll get it then."

Mom turned toward me and chimed in. "You should take it now, honey, while you have your car. Just put it in the trunk before you go."

"Okay, sure. See you at Christmas."

I said I would pick up the box, but I didn't. I would come back for Christmas and take it then. Everything would work out fine.

From my hotel room, I had a clear view of the hospital across the parking lot. I counted five windows in from the northeast corner on the third-floor cancer unit. Was that his room? The one with the dim yellow light of a reading lamp in the corner and the curtain halfway closed? I slept a few restless hours and woke up at four o'clock in the morning with the desperate urge to get back to my safe, normal work routines in New York. Part of me wanted to walk right into that cancer unit and wake him up to have breakfast, but it was too early. He needed to rest.

I packed my bag and got in the car with Caper. We drove to the stoplight at the edge of town, where the corner shops and cul-de-sacs turned back into the wild prairie. I sat idling at the red light. My father had always been the tough guy, the man who spoke simple truths and kept his mouth shut when he knew better. I had always listened to him, never the other way around, and he had said *I'll be fine*. I believed him. Like my narrow escape on the railroad tracks decades before, the good Lord was not ready to call him home yet. It wasn't his time. I understood little about where I was going or what leaving that day might cost me. The stoplight turned green and the frozen fields of South Dakota unspooled in my rearview mirror.

CHAPTER 6

FEAST OR FAMINE

I was back in the office the next week, sleep deprived, with a million things to do: expense reports, cash flow projections, and press interviews about First Crush. Although our recent wine sales had been strong, the business had yet to turn a profit that year and my stomach was tied up in knots. I weighed all aspects of the winery's finances and tried every scenario possible for making ends meet, but all roads led to the same conclusion: layoffs were necessary to right the ship. The employees were not anonymous faces in a crowd, but people I had worked with for a decade. I was an agriculture guy, someone who had a passion for wine, not a ruthless businessman accustomed to layoffs. Michael summoned me to his office in Manhattan to discuss the situation. When the drowning feeling started to tighten in my chest, I looked up from the rising tide of overdue invoices and saw myself having the same financially ruinous experience in agriculture that my father did.

THE FIRST TWELVE years of my life were shadowed by what is now labeled "the farm crisis." My father had leveraged our ranch's productivity as collateral to buy equipment and a foundation herd of cattle. I rode along in the combine as he harvested a bumper crop of wheat in 1987, which overflowed our granaries and piled up in a great bronze-colored mound in the yard. I would climb to the top of the hill and slide down on my butt, laughing, thinking it was my own private ski slope. The underlying problem was something far beyond what I could have understood at the time. There was a grain surplus caused by the Cold War embargo against the Soviet Union and my father could not sell his wheat. The following spring, the wheat mountain was still in the yard, causing a fermenting stench that attracted rodents and the rattlesnakes that preyed on them. My father took out loans to plant a mix of oats and barley, crops with more stable markets. The seeds sprouted tender and green in narrow rows as far as I could see—until the drought hit. We endured a fifty-day stretch without rain and on many of those days the temperature reached a hundred degrees. The crops burned to a crisp along with our alfalfa fields, so my father was forced to take out more loans to buy hay bales for the cattle.

"Galldang feast or famine round here," he used to say, kicking chunks of dirt that would disintegrate in the wind.

Failing or not, my father was working as hard as he could. Ranch chores started before dawn and did not end until after dark, or whenever the last birthing heifer was safely bedded down. He loved working the land not for the price of live weight steers per pound, but to experience the freedom of riding a horse across the prairie. He loved building fences and barns not to collect a paycheck, but for the satisfaction of seeing his own hard work turned into something useful for the family. While we never starved or worried about having a roof over our heads, I knew what it felt like to go without things for a lack of money. When friends from church visited, they pointed out the broken tractor in the yard with one missing tire; the pans of butchering scraps on the ground by the clothesline from which thirty-six cats were eating; and the clanking steel windmill that spit up gurgling brown water from the spout for five minutes before it ran clear. We were so far removed from any population centers that the hardships of ranching life did not seem out of the ordinary—it was simply all I knew.

After the drought, my father announced that he and Mom would have to take side jobs. Mom cleaned one-room schoolhouses in the area and came home in time for supper. My father helped us recite the Lord's Prayer before bedtime, then he put on a brown uniform and drove a UPS freight truck on the overnight shift to Denver. His body was aching and stiff by the time he got back the next day, but it was what he had to do. This routine went on for a few months until he made another announcement. He would be leaving for a welding job in a manufacturing plant a long drive east of us. That meant weeks away from home.

"Gotta move to town," he said. "Banks are taking the land."

LATER THAT WEEK, I drove to New York City with the financial analysis in hand, ready for my big meeting with Michael. I had planned to be in the city that day anyway, to have lunch with my Californian therapist friend, Dave. Coming around the last bend of the Long Island Expressway in Queens, with the sprawling Calvary Cemetery on my left, the Manhattan skyline suddenly popped into view. Every time, it never failed to take my breath away, the sight of those monstrous steel-and-glass towers looming over the East River, with people inside them making decisions that made the world go around.

THE FIRST TIME I had ever dreamed of living in New York City was the day my father took me to see *The Muppets Take Manhattan* in 1986, at the movie theater in Faith. It was the first movie I had ever watched in a theater, and Kermit the Frog exhibited a daring belief that anybody—even a frog, or a kid in cowboy boots—could show up in New York and make something of themselves. I was taught that Preszlers preferred the wide sky instead of skyscrapers, and mooing cows instead of honking taxis. Or perhaps it was not a conscious choice; we didn't know any different. *The Muppets Take Manhattan* changed my outlook. I began to believe something better might be out there, or at least something different from Faith. The economic collapse of our ancestral ranch deepened my resolve that if I had any smarts at all, one day I would leave and never look back.

I also sensed constant danger around Faith. Every cowboy and football player, every barber and bank teller that made eye contact with me for too long was aiming a loaded gun at my hidden sexuality. When I left the western plains, it was not so much for the intellectual enticement of job opportunities and cultural enlightenment. It was more like I heard the airhorn blast of an oncoming freight train, saying I had better get the hell out of dodge, and fast.

I enrolled at Iowa State University and immersed myself in its sprawling campus of twenty-five thousand students from a hundred countries. Anything felt possible. I suppressed my sexual desires in college by displacing them onto my botanical studies, a green safe zone that couldn't love me back but couldn't hurt me either. When the time came, my parents drove down to watch me deliver the student commencement address. Afterward, still wearing my graduation gown and mortarboard, I took only a few minutes to load up my dorm room into my little two-door hatchback—the one I had bought in high school after working a summer job scraping paint off a barn. My parents stood awkwardly for a moment after I loaded up the last box. I gave them each a brief hug.

"You get out there and do good," my father said.

I drove away, a couple weeks removed from my twenty-first birthday, and moved into a tiny, mouse-infested Manhattan studio with two roommates at the corner of Hudson Street and West Tenth. Greenwich Village in 1999 was not quiet—nowhere was quiet. My ears had grown so accustomed to Midwestern silence that I found it impossible to ignore the constant assault of city noises. There was the wailing of police sirens, the animated chatter of people strolling by on the sidewalk, and late-night carousing at the dive bar underneath my bedroom window. I heard every sound individually and thought I might go mad until I discovered foam earplugs.

My prospects began to change, and I tried never to look back. New

York City in the 2000s was full of marginalized gay men from rural areas who flocked here to find acceptance. We had moved along highways and vast stretches of America to the place we hoped would welcome us, where we might finally fit in. It did not seem so out of reach: the ability to map out a subway ride was not much different from the ability to navigate cornfields. You just had to know north from south.

Like the endangered species I had watched on *Wild Kingdom*, gay young men from the country learned to adapt in remarkable ways to survive. Earlier in my life, I had become an expert at hiding who I really was. In New York, I participated in everything and dove into the breach as though I belonged, until eventually I gained approval. In my twenties and thirties, I left my youthful struggles behind and escaped into the city's glamour, living a life that cycled endlessly between working and partying. At one point, I lived on the thirty-eighth floor of a high rise with a view of Times Square and the Empire State Building, just like Kermit the Frog. Having a skyline view was the surest sign possible that I had made it. My zip code told me I was successful.

I had tried to blend in like a chameleon with the energy, people, and money but I never did grow comfortable in New York's gay scene. I was still an outlier, almost two decades later. Despite living in the city of my childhood dreams and working at a posh vineyard, the bare truth was a bitter pill to swallow. I ran an unprofitable hobby farm for a wealthy movie mogul, while most gay guys climbed the corporate ladder in fashion, real estate, and media. They wintered in St. Barths and summered on Cape Cod while I commuted up and down the Long Island Expressway. It turned out that being gay was not enough for me to feel at home in New York.

Every change I made to appear more urbane and sophisticated on the outside just reminded me that deep down I was still my father's son. There was something else missing from my life. Whatever I did not have must have been profound; its absence left me feeling vacant and hopeless at times. I just could not put my finger on what, exactly,

I was lacking. The best way I had found to describe my predicament was that I was too gay for the country, but too country for the city.

Now, MY CAR crept to a halt in four lanes of expressway traffic waiting to enter the Midtown Tunnel. As I approached the tunnel entrance, my phone rang. The caller ID said *Home*. I answered.

"Hi, I'm going into the tunnel and will lose signal, can I call you back?"

There was a beat of silence on the other end before Mom spoke.

"Dad died."

CHAPTER 7

RIDING LESSONS

On a blazing hot afternoon, the summer after I turned eleven, my father took me horseback riding. I was of the age where I had taken up horses in a more serious way. Lucy and Mom were back at the house and my father and I were in the corrals. He stood next to my horse, Socks, grasping the leather reins to hold him steady while I climbed up a ladder and stepped into the saddle. Socks was an experienced and plodding horse, too old to get excited about much besides fresh oats. Even his galloping pace was so sluggish that Mom had cut a section of garden hose for me to use as a whip. My father's horse was a bay buckskin stallion named Chili that was young, skittish, and unpredictable—wound tight like a spring, ready to burst at any moment.

We rode side by side down the section line south of the house, a stretch of gravel with wheat on either side. Chili and Socks had travelled this route so many times that the language between horse and rider was relaxed. I dropped the knotted reins and reclined in the saddle with my hands folded behind my head. The sun had reached its peak in the sky, washing out the landscape in blinding white light. My father raised his binoculars at airplane vapor trails and I emulated him with my own tiny binoculars. I felt how small we were.

The gravel gave way to open prairie and we rode ten minutes to Black Horse Creek, which meandered through our ranch in oxbow curves. The horses sloshed through muddy, leech-infested, ankle-deep water. There was an embankment four feet high on the other side of the creek that Chili leapt up with ease. I gripped the saddle horn and squeezed my thighs on Socks before we scrambled up the slope, his hooves pumping and sliding as if he were climbing a greased stairway. After a few seconds, we were perched safely on the horizontal plateau, our hearts pounding.

The plateau was home to a stand of old-growth cottonwood trees that made no sense in the Dakota moonscape. They shot upward like fountains, with three or four trunks split from the same base in massive, contorted bouquets. We rode underneath them, and the temperature dropped twenty degrees in their shade. I glanced up at the cottonwoods, their leaves singing in the breeze.

I could pass a whole afternoon riding with my father and not hear him string together five words. I had been conditioned to think this was normal behavior. With the same deliberate repetition that he used to memorize Bible passages, he had taught me the names of prairie plants and wild animals. We trotted out from the cottonwood trees and a flock of sand-colored game birds glided above the grass some yards ahead.

"Sharp-tailed grouse!" I said, pointing.

"Yep."

There was a specialized grass dominating the prairie around us, with red-purple anthers dangling from the stems.

"Sideoats grama?" I guessed.

"Not that one," he said.

I leaned off the saddle by one stirrup to study the grass closer.

"Little bluestem?"

"I reckon so."

Socks trampled clumps of fan-shaped yellow flower buds on the broom snakeweed, with its taproot reaching twelve feet down to find water. Sagebrush and thistles quivered, bowing before every silent puff of wind. Yellow-breasted meadowlarks sang their flutelike calls from hidden nests on the ground. Locusts catapulted themselves and flickered their dry wings to flow around us, as though we were parting water. Occupying this vast space in isolation with my father gave me the feeling of independence that came from life on the prairie. Until that point, I felt safe.

Without warning, Chili reared up on his hind legs and gave out a tremendous bray. He thrust his head so high I feared that he might tumble over backward and crush my father. Instead, he landed hard on his front legs and bucked and kicked in circles. My father, being the seasoned rodeo professional that he was, weathered the fury and whipped Chili to remind him who was boss. The whipping made Chili buck harder, and in turn, my father kicked him with his boots, the silver spurs digging into his ribs. With the reins pulled taut and the bit in his mouth forcing his head high, Chili could no longer buck, and my father eased him forward into a wild but smooth gallop.

The panic in Chili caused panic in Socks. Before I could get my bearings, my normally comatose mount started bucking, too. Socks leapt wildly toward the cottonwoods, arching his back as he shot up and kicked, then smashed his hooves to the ground. My teeth bit down on my tongue and I tasted blood in my mouth. I could not focus on any one thing. The blue sky and brown prairie jumbled as if they were being shaken.

In all those summers riding horses with my father the most important advice he gave, which he repeated often, was this: never get your foot caught in the stirrup. If you did, you were likely to be dragged to your death or have your leg ripped off, whichever came first. I lost grip of the reins and my right foot slipped through the stirrup up to my calf. Hysteria set in.

Socks barreled toward the creek embankment and my instincts

screamed at me to let go of the saddle horn and pitch myself clear of pounding hooves before getting trampled in the mud. Socks reared up on his hind legs and sent me tumbling backward. Somehow, during the split second before I became fully airborne, a lifetime of advice from my father distilled into one desperate movement: I yanked my foot out of the stirrup. My legs flipped over my shoulders and I flew to the ground, sagebrush and cactus spines biting my back.

I lay there, stunned. The impact knocked every wisp of air out of my lungs. I struggled to inhale, to exhale, and to move. I was so disoriented that I felt, rather than heard, a rustling sound. Surely it was the cottonwood leaves blowing in the wind, right? My instincts told me something different: it was too sharp and high-pitched to be cottonwoods. The rustling grew into a full-blown rattle so primal that every hair on my body stood up in mortal terror. I turned my eyes without moving my head and saw what had caused the horses' hysteria.

A western diamondback rattlesnake was coiled up a few feet away. I lay there at its eye level, helpless. It tightened and drew its head back, its coil a thick rope of muscle prepared to strike if I moved. With the wind knocked out of me I already felt like I was drowning, but I held back the spasms in my lungs for fear of triggering the snake. Its tongue flicked in my direction. If its fangs sank into my face or neck, I would be unconscious within twenty minutes and in kidney failure within a few hours. The snake raised its head in a hideous hairpin curve, the last muscle position before a strike. I had to take in some oxygen. I couldn't hold my breath any longer.

A bullwhip cracked through the air and I flinched. In that tiny black gap when my brain lost sight of the snake behind my closed eyelids, its head snapped clean off. Blood splattered my shirt. The snake became a squirming headless mass of dry scales. My father pulled up beside me on Chili, holding Socks by the reins with one hand and his bullwhip in the other.

I looked up at the horses with their coats lathered in sweat, their bellies swelling and collapsing, their hooves pawing at the earth. My father dismounted. A rhythmic jingle of spurs approached with every footstep. He kicked the dead snake aside and stood over me in such a way that his cowboy hat cast a shadow over my face.

"How in the Sam Hill didya get yerself in this mess?" he said.

I lay there, gulping mouthfuls of air that could not make it into my lungs fast enough.

That was how I felt driving through the Midtown Tunnel in New York City, moments after learning of my father's death.

CHAPTER 8

WHITEOUT

The phone lost the signal and my car crawled forward in bumper-to-bumper traffic. I was trapped underneath thirty feet of bedrock, forty feet of clay, and ninety-five feet of water in the East River. All I could do was wait and waiting gave me a chance to catch my breath.

I emerged from the dim tunnel into blinding white sunlight and pulled over to the police zone. I shook my head at the timing of it all. I had been counting on our uneasy conflict to be resolved *before* his death. For all the time I had spent with my shoulders up near my ears, bracing for this hammer to drop, my long flinch was over.

In the world that existed a few minutes before Mom's call, I was planning to meet Michael and have lunch with my old friend Dave. I called Michael to cancel the meeting and he said, "Do whatever you have to do." I called Mom back and although she was talking, I only understood fragments, like I was listening to a staticky radio station: *went fast, hepatic coma, hospice, you don't have to drive, you should fly, it's too far, you work so hard, honey, don't beat yourself up, honey, it was in God's hands, how could you have known?*

She said the word *honey* in a specific tone that ended in a down note, the way she said it whenever I was upset about something as a kid. The single trait about my parents that exasperated me most was their acceptance of suffering. They were buoyed by an evangelical Christian optimism that if they submitted to Jesus, everything would turn out fine. Often, things did not turn out fine. I felt sick. I had to move. I had to get out of the city.

I drove back through the tunnel and called Dave to tell him what happened. I insisted that we still go to the restaurant. Of course, I was going to meet him for lunch! We had a reservation! We hadn't seen each other in ages; I wanted to catch up; I had to eat anyway; life had to go on. It all seemed so logical. I was clearly out of my mind.

I found Dave at Putnam's Pub & Cooker in Fort Greene, Brooklyn. He gave me a hug. I was too shocked to cry or speak. The waiter approached the table and asked an innocent question in a heavy Brooklyn accent: *How you doin'?* I had asked people *How are you?* countless times in my life—always with casual indifference, like the answer wasn't the point. I only said it to be polite. Now, hearing *How you doin'?* from a total stranger an hour after my father died felt like the air had been sucked out of the room.

The menu was indecipherable. Dave spoke with the serene voice of a seasoned therapist. "Hi. We'll have two cheeseburgers and the check. Thanks."

The food came and we didn't talk. Dave watched me eat and nodded his head with an approving kind of sincerity. His eyebrows arched in the empathetic way that people look at a dog after accidentally stepping on its foot. Dave understood that silence in the presence of a friend gave me comfort. Watching and seeing, but not doing or saying: this gesture had a simple name. It was called showing up. In the absence of affection, spoken or not, showing up was the way I had been conditioned to intuit love.

As I had done before many times in life, I operated on gut instinct. I told Dave I would go home to get Caper and then set out to drive two thousand miles back to South Dakota. He asked the obvious question.

"Why don't you just fly?"

I explained that my father told me on his hospital bed that he had left something for me in the garage, and I had to haul it back in my car. I wouldn't check it in on a flight or have it shipped. Perhaps the second cross-country drive was a way of punishing myself for not listening to him the first time. Maybe I just needed to get on the road and clear my head, and, anyway, I couldn't fly with Caper. Whatever the reasons, subconscious or not, I drove.

Mom's phone call was a dividing line in my life, just as certain and true as the dividing lines between the states that I drove across. I did not notice much of the scenery. My eyes barely registered the patchwork of frozen cornfields or highway grids or small-town truck stops. Instead I saw my father, his pained expression, an oxygen tube in his nostrils, IVs hanging from his arms. When I had left the hospital, I glanced back and saw him watching me walk away, his shoulders slumped, his mouth slackened. He was still holding that posture when I turned the corner. The look on his face would always haunt me: that expression of fear and loss.

One vital thing differed about my second trip across America: I turned off my phone. I stopped replying to emails. I didn't listen to the radio. It was a fully unplugged, electronic shutdown. The only sound was my car engine purring while it sipped diesel as yellow highway lines sailed past. No matter how hard I tried to focus in the silence, I could not arrange my thoughts in any meaningful order. My father was dead. That was as far as I got. My mind grabbed at anything it could, trying to make sense of the world. Walking Caper at a gas station in Toledo, I read my horoscope in a local paper, and believed it.

Taurus: During the day there will be minor troubles. You may have difficulty contacting people from afar.

Sometimes I sensed my father sitting in the back seat next to Caper as I drove. I checked my rearview mirror just to be sure he was not there, and then reassured myself that hallucinations were part of the ordinary madness of grief.

I was at a Best Western in Gary, Indiana, trying to get a little sleep when I dreamed about the cancer that I had not seen. I dreamed it was me in that hospital bed, my own body tangled in IV tubes, Mom cooling my forehead with a washcloth. I woke up sweating.

At the border between Iowa and South Dakota, I drove into a whiteout. The wind whipped the falling snow so violently that I couldn't see the road ahead, the wheat fields, or any farmhouses. One of my high school classmates had died during a whiteout after skidding into a ditch. He left his car to find help at a nearby farmhouse and got disoriented in the snow. The police found him tangled in a barbed wire fence two days later. After that, my parents stocked my car with tubs of peanut butter, beeswax candles, matches, and chunky knitted afghans.

"During a blizzard, the safest place is inside your car," my father said. "What you can't see can kill you."

After I don't even know how many miles, I was in a head-spinning, slow-motion vertigo and started to feel nauseous. I pulled over in the whirling and blinding snow and got out of the car. Instantly, I became disoriented and held on to the door handle like my life depended on it. I tried to vomit, but nothing came up. I dry heaved until my ribs ached. I felt as bad as I ever had. Heeding my father's advice from twenty years ago, I got behind the steering wheel and shut the door on the raging blizzard, hoping it would stop in a few minutes. It didn't. I couldn't see anything except streaking, billowing snow outside the windshield. I wasn't prepared with beeswax candles or peanut butter, either. All I had was a case of wine, a pedigreed dog, and one of Mom's quilts. I crawled into the back seat to snuggle with Caper and wait out the storm, hypnotized and terrified by the whiteout.

My father had to get creative around the ranch to have fun after all the blizzards. There was plenty of snow but no hills, so sledding was out of the question. When I was nine and Lucy was eleven, we discovered under the Christmas tree a wooden toboggan with a giant red bow made of hay baling twine. My father had built the toboggan out of ash planks that he steam-bent with his cattle branding torch and secured with rusty tractor bolts.

Somewhere between a broken vacuum cleaner and a deep freezer in the basement, Mom found the insulated snow coveralls that we dug out every winter. I put them on and ran outside with the toboggan, where my father was waiting with his horse, Chili. I held the toboggan while he tied his rodeo rope in a knot around the front handle. Mom emerged from the house wearing a full-length red velour bathrobe and leather moccasins. Her hair was rolled in baby blue curlers and her plastic eyeglasses made her look like a slant-eyed cat. She was carrying Lucy. My father stomped across the snowy brick patio in his boots and gently lifted Lucy out of Mom's arms. He sat her down in front of me on the toboggan. Lucy's legs were in molded plastic braces below the knee, and she wore a rubber helmet with holes in it, tufts of blond hair sticking out. He slid me forward on the toboggan a bit until my legs were wrapped around Lucy enough to keep both of us in place.

"Now look here, Trent, you've got two things to remember," my father said.

I looked up at him; his broad shoulders blocked the sun. He wore beige coveralls with motor oil stains and a gold zipper that ran from his waist to his collar. The knees and elbows were patched with squares of material that Mom had cut from red paisley handkerchiefs. Loose threads from the frayed ends of his pant legs draped over his cowboy boots, which he had polished and buffed to a black luster.

"Keep your eyes on me, in case I signal," he said.

"Okay, what's the second thing?"

"No matter what happens, even if this galldanged thing flips over, don't let go of yer sister."

He roped the toboggan to the saddle horn and swung himself up onto Chili. The sky was crystal clear and blue, and the temperature hovered around zero. Air gushed from Chili's soft pink nostrils in long, slow breaths that evaporated in white clouds. It was so cold that the night before, on Christmas Eve, I couldn't sleep because my father was in and out of the house chopping ice on the cattle drinking troughs with his axe.

With a light jingle of my father's spurs, Chili dragged us past the windmill toward the one-room schoolhouse. Lucy laughed in front of me as she pulled my legs up around hers. Our thick coveralls made rustling noises when we moved, and the snow squeaked underneath the toboggan. My father whistled and the toboggan shifted forward, faster.

"Hold on to yer butts back there!" he yelled.

Lucy squealed in delight while I laughed in nervous joy. Chili turned south at the section line by the bull pasture and picked up speed. Within seconds, the toboggan was floating on the snow-covered wheat field that extended south from our house, past dozens of haystacks half-buried in snowdrifts, and down into the dried-up creek bed with the old-growth cottonwood trees.

"H'yaw! H'yaw!" My father's voice carried a long way on the flat landscape.

Riding a toboggan being pulled by a horse was more fun than I could ever have imagined. Lucy and I were wedged so tightly that the biting wind could be avoided by dipping our heads into each other like two arctic penguins huddled against a storm. The rope between the toboggan and Chili grew slack as my father guided him around the cottonwood trees with the rhythm and grace of a slalom skier. To keep from falling off, we shifted our centers of gravity with every turn. The toboggan tilted sharply as it crested the snowdrifts and hugged the curves around the trees. I whispered a prayer that God would keep us from flipping over.

"Git ready!" my father yelled as we broke past the cottonwoods onto the vast open prairie. Chili accelerated into a full gallop. I pressed my face against Lucy's back as my father sped us through dips and around sagebrush clumps that were as familiar as the back of my hand. The only sounds were the rhythmic thud of Chili's hooves and the whooshing of snow grinding under wood. We were alone out there for a while, three souls connected by my father's rodeo rope. My heart was racing, and I held on tight to Lucy. She broke out laughing, a sound I rarely heard. This must have been what sailing felt like.

We were pulled in a wide arc back to the homestead. Chili slowed to a walking pace, and Mom emerged from the house with a Coleman thermos of hot chocolate. When it was safe to relax my grip, I let go of Lucy with a sigh of relief. She turned around and smiled; I smiled back. I was certain that horse tobogganing had to be the most exhilarating activity to ever happen in South Dakota. Judging by the ecstatic look on Lucy's face, she would have agreed.

My father dismounted and untied the rope. Mom poured us mugs of cocoa with tiny marshmallows floating on top. Christmas carols wafted from the record player inside the house.

"How was that, Lucinda?" my father said, smiling from ear to ear as he coiled the rope around his elbow.

Lucy would normally have stuttered and struggled to form words but did not hesitate this time: "FUN!"

People in town paid for those types of experiences, like drinking mulled cider on hayrack rides through a pumpkin patch. I got the riskier and rowdier cowboy version. The toboggan was the best Christmas gift, but I was equally grateful to have one pure moment with Lucy when she wasn't in pain, and we were both happy.

AFTER THE STORM cleared, I drove to my ancestral ranch. I felt a strong compulsion to stop and see the place, even though I needed to get back for the funeral. A huge rusted iron arrow that my father had welded together sometime in the 1970s pointed down a gravel road with no name. I drove the twelve-mile stretch back to the ranch that I had left behind decades ago, except it was not a ranch anymore. It was a monument to corporate agriculture. The new owners abandoned every wooden outbuilding my father had built and left them to rot in the ground. Across the road, they had built a shiny new homestead: big, perfectly square boxes constructed with red corrugated steel. Whereas my father had one Depression-era tractor, the new owners had six brand-new tractors, three combines, and nine trucks parked in a row next to the granaries. By the corrals where my father had worked cattle on horseback, the new owners lined up a fleet of four-wheel all-terrain vehicles. The ranch looked like progress done the new American way: invest and capitalize until it is too big to fail. I had heard rumors over the years that McDonald's Corporation bought up ranches in the area at rock-bottom prices, to create a vertical supply chain for beef hamburgers. If our old ranch were one of them it would have been an ironic twist, given that it was located farther away from a McDonald's than any other ranch in the country.

I parked halfway up the driveway, tires crunching across frozen gravel. When I opened my car door, the ranch that I once knew no longer existed. The sounds and odors of cattle I expected had been stifled inside giant industrial feedlots. With Caper trotting alongside, I walked into the bull pasture toward the abandoned one-room schoolhouse, forgetting that the pasture was no longer mine to walk across and no longer had bulls in it. I turned back toward my childhood home and waved my arms as if to reassure anyone inside the house who might be watching. It didn't matter; the house was abandoned. Siding pulled away from the walls, revealing tar paper underneath. On the north side, half a length of gutter hung down and the glass patio doors had been replaced with plywood.

I crossed the weedy patch that used to be a lawn and walked up to the kitchen windows. They were icy, lusterless, and dark. The wood around the windowpanes was rotten enough for me to scrape it away with my thumbnail. Cupping my hands around my eyes, I looked inside. The living room where we had once gathered around the Christmas tree was empty, with a few scattered metal basins. The oak trim that my father nailed to every doorway in the house had been stripped off. It wasn't clear how I should proceed. It crossed my mind to break into the house, but I didn't.

I walked across the expanse of frozen mud between the backyard and the schoolhouse. The swing set that Lucy and I raced around had sunk into the ground halfway up the poles. Dilapidated outbuildings slumped on ground that felt both familiar and foreign. The granary had partially collapsed, though my old basketball hoop still stood next to it, without a net. The chicken coop was gone. Egg-laying bins littered the ground. The corral I helped my father build as a practice rodeo arena was gone, the posts uprooted and stacked in a burn pile. The main barn where my father once stabled our horses was crumbling onto itself and caked with thousands of empty mud swallow nests. Inside the barn, sharp rays of light burst from cracks in the weathered gray wood walls. Still protruding from the wood was the nail where my father used to hang up the toboggan, but of course the toboggan was long gone. The tack room where we used to hang Socks's bridle and Chili's saddle was empty, save for the black outline of the Preszler ranch cattle branding iron that was burned into the wall.

At the schoolhouse, I sat where my desk used to be, near the stage my father had built for us kids to sing Christmas carols. A raccoon had made a nest there. Moldering encyclopedias and books were splayed on the floor under the broken south window, with air flowing through unabated. I picked up a book. It was a copy of *Where the Red Fern Grows*. Mice had chewed away some of the pages.

All these places were the waypoints of my young world and had magical importance, a pull on me, glowing with memory and meaning, however devoid of life they were now.

Not everything was dead and decaying. Standing alone in the center of the yard, thirty feet tall with a twisted trunk and dark green needles, a ponderosa pine tree stretched its crown to the sky. The upward-wavering twigs reached for the gray December light. Branches forked and clicked in the dry air above my head. I rubbed pine sap between my fingers to warm it up so I could smell that familiar scent. Eyes shut, I summoned the image of that Arbor Day years ago, when my father dug the hole to plant this sapling in its delicate tar paper pot. I had waited a lifetime for proof and now I had it. He was right. The tree had grown bigger than me.

CHAPTER 9

THANK YOU FOR YOUR SERVICE

It took five days to drive home in the snowy conditions. By the time I left the ranch and dropped off Caper at my parents' house, I arrived at the funeral home ten minutes late. Mom was already there greeting people at the door. She looked pale and worn.

"Where were you? I've been worried sick."

She leaned toward me and slumped her shoulders into my chest. I put my arms around her and held her while she cried. It might have been just the second time I had seen her cry. Her slight frame clung to me like a dried cornstalk. As she shook, my heart pounded. Holding her, I felt the difference between how much love we had stored up inside ourselves and how little we had expressed to each other.

"Mom, I—" The lump in my throat prevented me from talking.

Friends and family poured into the foyer, stomping snow off their boots and wrestling their winter parkas onto hangers. A hunting buddy of my father's, a gruff former bull rider named Dwayne, gave me a tight bear hug.

"Your father was a good man, the best kind of man," Dwayne said.

My father's friends were not the most effusive bunch, so hearing this surprised me.

Mom led me into the family visitation room next to the chapel. There was a wreath of red and yellow carnations, the colors of Iowa State, sent by the alumni president. Another wreath was made of miniature purple Minnesota Vikings football helmets.

"He still rooted for them even though they always lost," Mom said.

On a folding table with photos spread across it, there was a shot I had never seen before of a stylish square-jawed beefcake and a woman wearing big white sunglasses. They looked happy, relaxed smiles on their faces while they rode a wooden sailboat at sunset. I pictured them at a hotel on the beach for a romantic getaway and had the sensation of not being able to compute what I was looking at. The people in the photo could not possibly have been my parents. I put my nose a few inches away and stared at it, looking for clues that it was, in fact, them.

"That was our honeymoon in Hawaii," Mom said.

My uncle Floyd was there—my father's younger brother and business partner in the ranching operation. I always thought I resembled his two sons, my cousins, both in the military now and stationed on aircraft carriers somewhere. We had the same pointed Preszler eyebrows and full lower lips. The summer before I was in fourth grade—around the same time that the ranch was in a financial crisis—Floyd's sons beat me up after school. I stumbled home, bruised and bloodied. My father knew right away what had happened. It was not a hard mystery to solve, given that the school had only eight students. We moved away before classes started up that fall, so I didn't see or speak to Floyd's family again. Now, at the funeral home twenty-five years later, Floyd took off his cowboy hat and walked toward me. I turned my back to him, with no interest in platitudes.

There was a photo of my father wearing a coyote pelt hat, his head tilted back and his eyes squinted shut in laughter. He looked euphoric, like someone had just told him the world's funniest joke. My lone cousin from Mom's side of the family stood beside me looking at the same photo.

"That's the hat he brought to Christmas dinner a few years ago," she said.

"Oh, you guys spent Christmases together?" I asked, surprised.

I had not come home for so many years that I had scant understanding of my father's relationships with other people. More cousins, aunts, uncles, and church parishioners whom I had not seen in ages walked into the room. Some had been married or divorced without my knowledge and stood next to spouses I hadn't met. Some had kids whose names I didn't know and faces I didn't recognize. They introduced themselves with halting formality, like I was a stranger. I felt awkward and out of place, an interloper at my own father's funeral. I had no idea where to stand, what to do, where to place my hands, or how to manage my facial expressions when people looked at me—and people were looking at me, monitoring how I reacted to the setting and the conversations. I was self-conscious, afraid of being judged as the gay prodigal son.

I began feeling clammy, so I stepped outside to cool off. The mortician, who was also my high school band teacher, came outside and regarded me with a sad smile. He walked me back into the viewing room where my father's body lay in an open casket.

He could have been a dressed-up wax mannequin, but when I saw the hands folded across the chest—*our* hands—I knew it was my father. His face was clean-shaven, and his jaundiced skin was dabbed with thick makeup. When he was shaving in his hospital room, maybe he knew it was his last chance to clean up. He wore the same navy blazer that looked too big on me in my high school senior photo. The green necktie with embroidered ducks that I gave him for Christmas when I was a kid was tied in a Windsor knot around his sagging neck. I asked the mortician-slash–band teacher to remove the duck tie before my father was buried, so I could keep it.

I leaned over the casket and stared at him for quite a while, ignoring the shuffling of funeral guests filing into the chairs behind me.

When I had said *See you at Christmas* in his hospital room barely two weeks ago, I was making a clear-eyed statement of fact. It did not occur to me that he could die so soon after I left. For so much of my life, he was monolithic. Perhaps I didn't expect I would have a deep relationship with him after our falling-out, but even if I didn't factor into his daily life, I still thought he would grow old—that he would go duck hunting on Saturdays and watch football on Sundays. There was supposed to be plenty more time to have the conversations with him that for thirty-seven years I could not.

I had seen movies where people stood over caskets and spoke about abiding love or apologized for things, so they could go on living. I could not bring myself to do even that. I just stared at the man who sat by, blank and unreachable, while I worked to get ahead in the world. I had lived in a near-constant state of longing for the father right in front of me yet out of reach. What I wanted, and what my life lacked, would soon be buried along with his corpse.

The minister entered the mortuary chapel wearing a black robe that covered all six feet plus of him. He spoke with a booming voice and gazed across the tops of our heads toward the back wall. The parishioners and family members all knew when to stand and sit and sing during the liturgy. I did not. The minister spewed melodramatic threats of hellfire and brimstone during his sermon: *Doth ye not repent to Jesus, ye shall burn in the eternal flames of hell.* It was a familiar screed from childhood, this horrific fear mongering of the consequences of sin. Even after all this time, people still said and believed these things.

The minister went off script for a few minutes. He spoke of my father's love for the outdoors and hunting, for woodworking and craft, and expressed appreciation that my father helped make repairs around the church when money was tight.

"He even built the altar," the minister said. "But above all else, he loved his family."

My head swiveled around the room. People nodded in agreement. I could not believe my ears. He *loved* his family? If that love included me, he never said so. Did that mean he told the minister and his friends that he loved them, even though he never actually spoke those words aloud to me? Maybe the minister meant to use the l-word as something generic, intangible, and out of reach—like *For God so loved the world.*

Mom didn't ask anyone to deliver a eulogy. I didn't question how she arrived at that decision or why she didn't ask me to speak—Lord knows, I would not have wanted to anyway. I ran down the list of all the family funerals I had attended, and none included a eulogy. My father had always said that Lutherans weren't big talkers.

The service was over in a flash. The minister announced that the burial would happen at the Hills of Rest Cemetery, ninety miles north. Pallbearers loaded my father's casket into a hearse that was parked in front of the funeral home. A police car with red flashing lights idled nearby, ready to escort us out of town. The exhaust fumes from both vehicles billowed into the freezing air and enveloped everyone standing underneath the carport awning in a smokey haze. I studied Mom's face. She cast her eyes downward, lost in thought. Fear was etched into the quivering of her lips and the creasing of her forehead. She did not have a private persona for me and public persona for others. She was the same Mom I had always known, somehow managing to appear fragile and gutsy in equal measure. The pallbearers closed the hearse's door.

On the ride, Mom and I sat beside each other in the back seat. The funeral procession crept through a white expanse of cattle pastures and cornfields so endless that it seemed we were taking my father to be buried on the moon. When we encountered such a procession as a kid, I would comment that all the cars in the oncoming lane had their headlights on. My father would turn on Old Yeller's headlights and turn off the radio until every car had passed.

Somewhere along the route, another police car sat with its lights flashing. An officer emerged and saluted as we drove past. A few miles later, a cluster of Vietnam vets stood at attention roadside, holding

American flags. The outpouring of appreciation from uniformed officers caught me off guard. Mom had not mentioned anything about a military funeral, though my father had served in Vietnam.

"Well, isn't that nice of them," Mom said, removing her glasses to wipe tears from her eyes. "That reminds me, I need your help with the Agent Orange forms back at the house."

"Huh? What's that about?"

She hesitated and took a long, drawn-out breath.

"The doctors think his cancer was caused by Agent Orange exposure in Vietnam. There's a survivor benefit I'm supposed to apply for."

"Oh God. Wasn't Dad a sniper up in the trees or something? He must've been doused with that shit every day. And now the government wants to pay you for the inconvenience?"

For the remainder of the ride to the cemetery, I researched Agent Orange on my phone and grew enraged.

"Says here 2.8 million vets were exposed," I said, reading aloud from my screen. "Looks like you'll get a hundred bucks a month—infuriating!"

"Honey, don't even go there today," Mom said, her voice wrecked by a heartbreaking sob.

I scanned the list of diseases on the Veterans Affairs website that are now known to be caused by Agent Orange. Not surprisingly, my father's form of cancer made the list. Then I clicked on a bulletin from a federal scientific agency that described a link between Agent Orange and birth defects in the children of troops who returned from the war. My heart sank. Lucy's disease was listed in the same class of genetic mutations caused by the nerve toxin that was likely passed down to her from my father. I didn't share this awful information with Mom. I slipped the phone back in my chest pocket and stared out the window. We rode the rest of the way in silence.

When the procession pulled into the cemetery and Mom stepped out of the car, a row of military officers saluted the hearse. They wore gold-braided cords around their shoulders and clinking medals on their chests. The officers, not the pallbearers, carried my father up a gentle slope to a hole in the ground with a pile of dirt next to it. Mom and I followed along with about a dozen family members. Our feet crunched through a thin layer of ice into the snow beneath. There were no headstones, only in-ground grave markers concealed by snow. The cemetery spread across a rolling meadow with a clump of Colorado blue spruce trees growing in the center. My father's casket rested in the shadow of the trees. Their crowns were laden with cones and heavy snow, bending each branch toward the ground. I saw in the spruces, as I did with all trees, their complex geometry of intermeshed cells coursing with life beneath cracked bark. Stored energy from their roots fountained up into a canopy like a big elevator system for life. They had grown quite a bit taller since the last time I was here, fourteen years ago, on the day we buried my sister Lucy.

If fundamentalist Christians could believe in such unscientific notions as heaven and hell, then I could believe that the spruce trees were pouring out messages to me through their branches and needles. What they said to me was: they had recorded in their wood the history of every life that had been buried near their roots, including Lucy's. I thanked the trees for keeping her company all these years.

"Are we burying Dad next to Lucinda?" I asked, always using her full name around my parents. Mom looked away, too distraught to speak.

A few feet to the left of the freshly dug hole, I got down on my hands and knees and brushed aside the snow to reveal a rectangular bronze plaque. I placed my shivering, wet hands on its surface and felt the engraved letters that spelled her name: Lucinda Mae Preszler.

We slumped into cold metal folding chairs next to the casket. Small ice particles my father used to call diamond dust sparkled in the windless sky. The minister spoke about sinning, repentance, condemnation, and resurrection. My eyes drifted toward a billboard in the distance with a Jiffy Lube motor oil advertisement and one of those signs that repeatedly flashed the time and temperature in red lights. It hypnotized me: *12:32 PM*, *9°F; 12:32PM*, *9°F; 12:32 PM*, *9°F.* I was snapped out of my trance by the sound of a sergeant calling to arms and his comrade firing three gunshots into the air.

Ready, aim, fire.
Ready, aim, fire.
Ready, aim, fire.

I flinched with each shot, but the smell of gunpowder was a comforting memento of childhood days spent hunting with my father.

We endured a long and awkward silence while two officers folded the American flag that had been draped over the casket. They folded it wrong and had to refold it—twice. Each time took an eternity while they fumbled with tucking in the last corner of fabric. On the third attempt, they got it right. One of the officers knelt in front of Mom and presented her with the folded flag, "On behalf of a grateful nation . . ." She reached for it haltingly, as if she thought, *Oh, I'm supposed to hold this thing now*. The symbolism of the flag was meant to give her comfort but clearly had the opposite effect. I put my hand on hers and touched the embroidered white stars on the flag's blue backdrop. The pained look on Mom's face seemed to say, *How did we get here?* I hoped I didn't have the same expression on my face. I tried to be strong for her while hiding my surprise at this full military burial. I had obviously missed some crucial details of my father's life that merited such pageantry. Likewise, I was sad that he had missed so much of my life.

An officer stepped underneath the snow-laden boughs of the spruce trees, between the graves of my father and sister. He turned to face the western sky and raised a bugle to his lips. The silence left that snowy hillside like it had dropped through a trapdoor. His first note, B-flat, was clear, certain, and loud. It pierced the frigid air. He held the note a long time, but the second note was cut short and gone too soon. The last note of the first phrase rose to an untouchable place in the sky where it vibrated over the treetops.

Sitting in that cold metal chair listening to taps, I had questions for my father that I could never ask. Who was he to me, and who was I to him? Did he really love me like the minister said? Had he progressed at all in understanding his son? I felt at a distance from my father, as if he were a foreign country. I did not speak his language or become steeped in his customs, and I had no passport to travel in his country. I ached that I wouldn't be able to unlock the door leading to everything I didn't know. The final note of taps echoed across the prairie. When that note faded off, any hope I had of knowing my father disappeared with it.

CHAPTER 10

HOUSE OF SECRETS

The next day, I helped Mom fill out the Agent Orange death benefit survivor forms, which required identification numbers from my father's army discharge papers. She brought up from the basement a heavily taped shoe box that she said contained the paperwork we needed. She described the day my father had come home from Vietnam in 1970. He had dumped out the contents of his olive green army-issued duffel bag and saved his paperwork along with a few small personal items. After burning the duffel bag in the barnyard, he sealed up his papers in this shoe box, put it on a shelf in the basement, and asked Mom not to open it until he was gone. She had almost forgotten it existed until now.

I slid a knife under the taped edge and opened it. A small black jewelry box was sitting right on top. It was heavy, wrapped in supple leather with a decorative gold pattern around the edge.

"What's this?"

"I don't know, never saw it before," Mom said, leaning forward with a confused look.

Inside was a small metal star dangling from a red ribbon. A slip of paper was tucked into the lid. I unfolded it. The words of the document were printed in bold letters. A gold embossed emblem of the U.S. government was set at the top of the page and the signature of the secretary of the army was at the bottom, dated on the eighteenth of August in 1970. I read it aloud to Mom:

THIS IS TO CERTIFY THAT
THE PRESIDENT OF THE UNITED STATES OF
AMERICA AUTHORIZED BY EXECUTIVE ORDER
HAS AWARDED
THE BRONZE STAR MEDAL
TO
STAFF SERGEANT LEON K. PRESZLER
FOR
MERITORIOUS ACHIEVEMENT
IN GROUND OPERATIONS AGAINST HOSTILE FORCES

His rapid assessment and solution of numerous problems inherent in a combat environment greatly enhanced the allied effectiveness against a determined and aggressive enemy. Despite many adversaries, he invariably performed his duties in a resolute and efficient manner. He energetically applied himself to each task which contributed immeasurably to the United States mission. His loyalty, diligence, and devotion to duty are in keeping with the highest tradition of the military service and reflect great credit upon himself, his unit, and the United States Army.

I touched the gold seal with my fingertips and read the words a second time.

"I guess this explains the military burial," I said.

"I guess so."

"Wait, you didn't know about this?"

"I had no idea. This is the first I've seen of it. Dad didn't want to talk about the war, so we didn't."

I imagined him in the thin light of a lamp somewhere, reading the commendation, folding it, tucking it into the little black case, and vowing never to speak of it again. If he had not told Mom, then he had good reasons.

"Maybe he wasn't proud of what he did to get this," I said.

A small photo album in the box helped add some color to a story we knew nothing about. It included images of his battalion moving through the jungle, boarding a helicopter, and traversing a river holding guns above their heads. In one photo, my father appeared to be showing Vietnamese children how to use a rodeo rope. He had constructed a makeshift dummy steer using ammo boxes and sandbags.

As I rifled through the box, something else caught my eye: security clearance papers and military travel permits for Cambodia. I scanned over key words like *Special Forces* and *Top Secret*. Page three of his discharge included a signed affidavit that he would not reveal secret government information during the rest of his life. There was his unmistakable signature in black ink next to a checked box that read "Extent of Access, Top Secret."

I wondered if I now had three fathers: the one I did not know at all; the one I knew a little but wished I had known better; and the one who knew I could never truly know him, no matter how much I wanted to.

"So, let me get this straight," I said. "The whole time the shit was going down with Lucinda and the ranch, Dad was keeping *this* bottled up inside, too?"

"His mission was top secret, so he kept the secret!" Mom said. "He was a man of his word. You should be proud."

"Still, you'd think he'd tell his wife, at least."

"Maybe you were both good at keeping secrets."

I shrugged off her comment and focused on polishing the Bronze Star with my shirttail. Despite her sadness and her quiet manner, Mom was focused and resolute. She asked me to write a letter to the U.S. National Archives, filing a next-of-kin request for access to my father's military personnel files. The archives' website said it could take between three and six months to hear back, but we might learn more about his service record and why he received the Bronze Star. I signed the letter with hope and posted it to Washington, D.C.

MOM ASKED ME to sort out my father's clothes. I opened the closet in their bedroom and separated shirts, coveralls, stocking caps, and jeans by color and weight.

"Don't feel like you *have* to take anything, though it *might* be nice," she said.

I tried on his black calfskin Tony Lama cowboy boots and a pair of Levi's 501s. I cinched them with a leather belt and his hefty buckle. The buckle was four inches across, about the size of a teacup saucer but oval, and must have weighed half a pound. It was a trophy, made in Montana by the pro rodeo cowboy association and featuring a design that had been in place for the last half century. The gold and sterling buckle had been cast from a mold with edges formed to resemble braided rope. There were two Montana rubies set in flowers cast from green gold with a filigree of rose gold. The center of the buckle depicted two cowboys silhouetted against the Rocky Mountains: my father on horseback roping a steer's horns, and my uncle Floyd roping its heels. No one could really place a value on the buckle; it was one of a kind, engraved with my father's name and the year of his championship, 1968.

The buckle was a source of both pride and ire for my father the instant it arrived with our last name misspelled as Pressler. He had a jeweler correct the typo, but when the buckle came back, the *Z* was a different colored gold alloy than the rest of the letters. Nobody got rich on the rodeo circuit, not even champions, after they paid the entry fees, hayed the horses, and gassed up the trucks. The name recognition and pride that came with a victory mattered a whole lot.

"Can't get my own danged name spelt right, what's the flippin' point?" my father said.

Trying on his clothes was a strangely intimate experience. I found an old hardware store receipt and loose change in the jeans pocket.

As I unfolded, tried on, and refolded his clothes, it became easier for me to picture him as a younger man: pulling on his spit-shined boots and threading his belt through the loops with his big-knuckled hands. On Friday night a few times a year, he and Mom would say goodbye and walk out dressed up for something—Mom with a fresh perm and lipstick, my father smelling like sandalwood cologne. By ten o'clock, they'd be home. My father would sit at the dining table and pour himself some whiskey while Mom drank a can of Tab, ice cubes jingling in tall glasses. Wearing my father's clothes, the years separating us disappeared for a moment and I became him, somehow.

I took his cowboy boots, belt buckle, and two pairs of jeans—the kind of worn denim from forty years ago that people in Brooklyn call vintage. It was hard to leave behind things that had been important to him but were now important to no one.

When the clothing job was done, Mom asked me to return my father's pills to the pharmacy. I started sorting out several dozen amber bottles from my parents' medicine cabinet, separating the painkillers from the hormones and chemo, but eventually, I gave up. I swept the whole mess into a plastic shopping bag that rattled as I moved through the house toward my car.

"Tell the pharmacist he's dead so they don't keep refilling his prescriptions and billing Medicare," Mom said on my way out the door.

A foot of snow had fallen overnight, leaving a smooth, untrammeled slope of white in the driveway. The December air hovered around five below zero. It was the dry cold that every Midwestern kid grew up knowing intimately: it froze the insides of my nose together when I breathed and caused ice crystals to form on my eyelashes. It was the kind of cold snap during which my father used to tell me never to lick the swing set or I would be stuck there till spring. By the time I finished shoveling the driveway, and drove to the pharmacy and back, I was frozen from head to toe. I went inside to warm up in the maroon leather recliner where I had last seen my father asleep with Caper at Thanksgiving. Next to the recliner, his heavy German binoculars sat on a stack of *Field & Stream* magazines. I picked up the binoculars and focused them outside. Snow blew across the fields and skimmed the railroad tracks. Wind took corn husks with it and piled them up in the corner of a barbed wire fence. A bird feeder swung from the sugar maple tree my father and I planted when I was in high school. Cardinals and nuthatches flitted around, crushing tiny seeds with their beaks. I put the binoculars around my neck and felt their weight swinging from the old leather strap.

Mom peeked her head into the living room. "Oh good, you're back. Follow me."

She led me through the garage, past the stacks of Tupperware filled with Christmas cookies, past pegboards and tools, a dismembered soil tiller, and a claw-foot oak table sprawled with its legs in the air. As on Noah's Ark, each species of suburban hardware was stored with its own kind: lawn chairs with lawn ornaments; garden shovels with leaf rakes; birdseed with bird feeders; snowblower and lawn mower with red jugs of gasoline standing in stiff ranks opposite. I edged past a table saw in the middle of it all.

At the back corner of the garage, a wall of crooked cabinets and smudged bar mirrors hung frame to frame over my father's workbench. Cardboard boxes and tin cans of all shapes and sizes were strewn around. Light from a single bulb in the rafters filtered down through racks of elk and deer antlers, fishing poles, and mysterious building jigs and contraptions I did not know the names of. It all seemed less like an installation in the corner of a garage than a creature under enchantment—like the whole bench might upend itself and trot away across the prairie.

My heart went still at the silence of this cluttered treasure corner, glowing with the light of an old saloon. A shut-up heaviness weighed over the space, the air stale and reeking of varnish, fertilizer, and beeswax.

"I've still got lots to sort out, and you probably don't want his guns, but this is for you," Mom said, pointing at a cardboard box on the benchtop.

My father had written my name on it. A duck that had been stuffed by a taxidermist sat on top of the box. Its feathers were covered in a thick layer of gray dust and cobwebs, but their beauty still shone through. The duck's beak was a flattened conical tube of blushed red and translucent gunmetal, with a jet-black tip like it had been dipped into an ink well. Emerald green feathers cascaded over its head, resembling an aerodynamic bicycle helmet. The breast plumage was scalloped in a riot of saffron and bronze, which grew into a dark chestnut color on its neck. Its eyes, those beady black eyes that I could never forget, were encircled by a thick ring of cherry red skin. I could have sworn its eyes followed my movements.

"All the hunting and fishing he did over the years, he must've killed thousands of animals, but the only one he ever had stuffed was that stupid duck," Mom said, shaking her head. "I always wondered, why the duck?"

"I was hunting with him that day. I've always loved that duck."

"Good, I wanted it gone anyway. I never did like taxidermy. Creeps me out."

My father had mounted the duck on slabs of weathered barn wood and meticulously arranged sprigs of marsh grasses around it, giving it the appearance of a diorama of exotic wildlife at a natural history museum.

"It'll look great in my house," I said.

"You should open this," Mom said, tapping the box with my name on it. "Dad was trying to get you to take it at Thanksgiving, but you were in a hurry."

I set the duck aside and untied the red twine holding the box together. The cardboard panels unfolded to reveal a wooden chest that was about two feet tall and three feet wide, with a million dents and scratches marring every surface.

It was my father's toolbox.

I ran my fingers along the scarred flanks of the sideboards, which had worn asymmetrically and chipped on the corners, like the whole thing had been dropped from great heights numerous times. There were four apothecary drawers in front with bronze ring pulls, but the carrying handles on the sides had busted off. The mahogany-colored wood it was made of had a flat, dead quality, lacking a certain glow—a patina that came from years of being tossed into trucks, hoisted into hay lofts, and kicked by bulls.

I untied the frayed twine holding together the chest's bronze latch. That opening of the lid and the moment of hearing the hinges squeak and of looking inside, then choosing an object, and reaching in and picking it up, was a moment of revelation. My hands trembled when they touched these objects, all charged with my father's energy. The aromas emanating from inside—leather, tobacco, grease, cedar—transported me back to childhood days in the 1980s, when I spent time with my father in his shop, watching him do slapdash repairs and rub sweat off his forehead with the back of his wrist. I could picture him like it was yesterday: opening his box of hidden things, fishing around inside, and taking tools out to be applied to some earnest task, like building a toboggan for Lucy and me.

"What am I supposed to do with this?" I asked.

"It's not up to me. Do what you want, but the main thing is just to keep it safe. Dad thought you might find a project. You know, he could build anything."

I closed the lid, wrapping the frayed twine around the latch with great care. To contemplate the life of this beat-up toolbox sank me into calm like a stone in deep water, so that when it was time to move it to my car, I walked out, blinded by the snow's glare, hardly knowing where I was. I buckled the duck into the front seat and sat the Bronze Star and binoculars next to it. The toolbox went in the back seat. I trudged through the snow into the garage, where Mom was waiting.

"So, that's it?" I said.

"Yes. That's your inheritance."

CHAPTER 11

DUCK HUNTING

The duck met its fate the last time I ever went hunting with my father. It happened around Thanksgiving in 1997, when I was home on break from my senior year at Iowa State. He had called a few weeks before, inviting me to go duck hunting in the fabled Lewis and Clark Lake cattail slough, near the spot where Sioux elders smoked a peace pipe with Lewis and Clark in 1804. I agreed to come home partly out of obligation to my father, and partly for a change of scenery. On the phone, he sounded the way he always had back in our less complicated life, when Lucy was still able to walk and talk and eat, and we still lived on the ranch, and we were more of a family. Something I had wondered about my father, who was fifty-two years old then, was: Did he want to go hunting to remember our life as it used to be?

We raised our bones at four o'clock in the morning and drove to the lake with the truck windows down to let in freezing air, to account for our sweating in thermal coveralls and camouflage hip waders. We arrived and walked for nearly an hour in the predawn darkness, navigating through dense cattails with headlamps strapped to our foreheads. The morning air felt heavy and the fog clung to my hands and face. I struggled to lift my feet through the muck, which was green-black, stuck to everything, and reeked of tar and decomposing vegetation. My father swayed under the weight of a canvas backpack of duck decoys that made him look like a giant turtle.

I carried both our shotguns, one slung over each shoulder. They were matching Winchester 12 gauges with walnut stocks my father made himself. He had split a walnut log and hung the halves in the garage to dry for two years before whittling them down to the shape of the gunstocks. He carved a checkerboard pattern into the grips using an old ice pick that he had heated in a fire and filed into a sharp point. The orange shellac and linseed oil that he boiled in a Crock-Pot filled the house with aromas as he sealed the walnut and sighed his way through umpteen coats, buffing the finish with an old T-shirt.

Deeper and deeper into the slough we went, until the path became almost impossible to follow. There was a clearing up ahead below the tangled branches of an ancient, leafless cottonwood tree.

"Look," I murmured, holding out my arm to stop my father. Something silver and green reflected off our headlamps from inside the clearing. "Is that a boat?"

He merely smiled, knowing it was there all along, but apparently wanting me to discover it for myself. We high-stepped through the water as the sun rose over the horizon and cast long shadows across the marsh. When we reached the clearing, my jaw dropped at the sight of an aluminum canoe bobbing up and down on the lake's rippling surface. I then saw that the canoe doubled as a grass-front hunting blind, built of wood slats driven into the mud and lashed with massive clumps of cattails for disguise. Chicken feed buckets were lined up inside the canoe for hunters to sit on and not be seen by flying waterfowl. I ran my fingers along the canoe's top railing, gripping the hollow pipe bolted across the middle. The muscled hull was held together by smooth rivets like those on an airplane wing. It was dented and rusty in spots, but I didn't care. Within seconds of seeing it, I let the canoe occupy a mystical place in my mind.

"Did you build all this?" I asked.

"Nope, it's Dwayne's blind, I just borrow it."

"You know, if you bought a bigger boat you could access the blind from the water and save the hassle of stomping through all that muck," I said with collegiate know-it-all confidence.

"Zip it, smart aleck. Can't afford no boat."

He tossed decoys into the water in front of the blind and they splashed down, their bobble heads popping upright as their lead anchors sank. He then held the canoe to stabilize it while I gingerly stepped in and sat on a feed bucket. When he slumped into the bow we nearly capsized under the force of his weight. The inside of the canoe, which was about seventeen feet long and four feet wide, was lined with dried reeds and littered with shotgun shell casings and empty beer cans. We propped our guns on our thighs and loaded three shells into the chamber. My father could fire a gun, take it apart, clean it, put it back together, and reload it with a blindfold over his eyes—a life skill he had learned in the army. I was not as hardcore as him but could still handle a 12 gauge.

We faced the silent, barely rippling surface of the lake, beyond which, through more cattails, was the rolling brown water of the Missouri River. We sat together in the soundless dawn waiting for the sun to rise high enough to burn off the fog. The canoe tilted radically to one side as my father reached into the decoy bag and pulled out a green Coleman thermos of hot cocoa for me, and a whiskey flask that he tucked in the dried reeds between his feet.

"How's Lucinda been doing?" I asked.

"Fine," he said, though I knew he was lying. In the poor light, his heavy brow and eyes looked black and penetrating.

Lucy was living full-time with a feeding catheter in her stomach at the children's hospital in Sioux Falls. Mom was well into her new career working the registrar's desk at the University of South Dakota, which also meant taking night and weekend classes in computer programming.

"Do you and Mom still take her out of the hospital for church every week?"

"Yup. She even sings a few words sometimes," he said. He didn't look up as he fiddled with the zippers and pockets on his hunting jacket.

"She always had a sweet little church voice," I said.

"She sure did," he said, before abruptly changing topics. He looked at me in a way that seemed to stare right through me to the morning sky above the duck blind that had turned to fire. "If you ain't goin' to church in Iowa, you ought to say your prayers now and then."

"I do."

He nodded and seemed to end the conversation: "Let's see about callin' some ducks."

The sun became a white burning disk behind the fog in a part of the slough opposite from where I first expected to see it. In the trek out to mid-water I lost my bearings and had no idea where we ended up. The disorientation felt exciting, though, like we had escaped to a secret foreign country, just the two of us. With the fog burned off, I could now make out the gray-brown open water of the lake, dotted with low, frozen grass islands. A bald eagle perched on the outstretched limb of the ancient cottonwood, black silhouettes against the sky.

Off in the distance across the lake, truck doors slammed, and boat engines puttered. We were near a shabby little state park with low-slung docks and permanent barbecues mounted on steel posts by campsites. Growing up, my father had brought me here to fish for bass using frogs that we caught ourselves as bait. Whenever we did hunt here, we would sit onshore waiting to pick off wayward birds that fled the hunters out in deep water, with their fancy motorboats. I had seen other kids in high school come here with their fathers: doctors and lawyers from town who had leased parts of the slough and built permanent duck blinds and heated ice fishing shacks. It was a famous place in the way that hunting and fishing camps can be mysterious and have a danger about them for young men. It was a place where grown men were dirty, cussed often, drank beer, smoked cigars, and urinated into the lake while standing on the docks in full view. Whenever we came here, I withdrew into a shell, becoming awkward and shy. There was some consolation, though, in being in the company of men who shared a passion for the outdoors, which made our differences seem not as threatening.

Even under perfect circumstances, there was always something at the lake that left me feeling unhappy, unlucky, or both: freezing fingers, muddy boots, the likelihood of getting skunked, which meant not catching any fish or shooting any birds.

"Quit yer belly aching and get used to bein' skunked," my father would always say, "so yer head'll be screwed on right when good stuff happens."

After a long while in which we saw nothing, a pair of ducks whizzed by out of range. When they had flown so far away that they were black specks in the sky, hunters in the distance took their shots, bursts of muffled gunfire in rapid sequence. The black specks fell straight down.

"They's wakin' up! Won't be long now," my father said.

He offered me his flask and I took a sip. The whiskey was so sour that my whole body shuddered, and my eyes watered when I swallowed. I looked at him to gauge his reaction; he scanned the milky sky, cool as a cucumber.

My father removed a hand-carved wooden duck call from the zippered pocket of his coat, took a deep breath, and lifted it to his lips. The call began with a long, high-pitched rasping followed by many short grunts and clucks, the sound of contented ducks dabbling in shallow water. Something came over him when he called, a preoccupying focus that removed him from my presence.

I crouched behind the wall of reeds and wood slats. Being so close to my father, I smelled whiskey and the Vicks VapoRub he smeared under his nose. After a few minutes, a flock of ducks flew overhead and turned on a dime. They veered toward our decoys as my father's calling intensified. The flock glided in low toward us with cupped wings. I clicked my gun's safety button to the left and slid my finger into the metal loop surrounding the trigger, all while keeping my eyes focused on the flying birds.

"Not yet," he shout-whispered at me from the corner of his mouth.

I shuffled my feet to get a base under me.

"Wait a second . . . okay, now!"

I stood up and the canoe lurched precariously sideways while I pulled the heavy Winchester into my right shoulder. The ducks briefly touched down on the water and erupted when I popped up from the blind. They climbed up and up, fast, flying almost straight backward away from us with their necks outstretched. I didn't really aim or even look, I just pulled the trigger on instinct, three times with a cock-back to empty the chamber of each spent shell. The smell of gunpowder burned my nose and a gray paper wad from the shell casing drifted overhead on the breeze. The flock changed direction and flew away, but two of the birds fell and floundered among the decoys. My father looked at me, raised his eyebrows, and smiled his pursed-lips smile.

Just then, the bald eagle that had been sitting in the cottonwood swooped down from its perch and snatched a duck from the water. In the same fluid movement, it returned to the tree, carrying the struggling duck in its talons. I was awestruck and giddy, having just witnessed the real-life version of a scene from *Wild Kingdom.*

My father was up and out of the blind in a hurry to retrieve the second bird. When he brought it back, I saw that it was a duck all right, but it was not dead. It held its head high and looked at us. This was unlike any bird I had ever seen, and not the typical mallard we would pick off from shore. My father smiled.

"Will wonders never cease, it's a wood duck," he said, shaking his head. "Never seen one before."

"What's it doing here?"

"Hard tellin' not knowin'. Could've joined up with those mallards for migration."

"What do we do with it now?" I asked, knowing full well what the answer would be. We had shot countless animals over the years, and any time something was wounded, we put it out of its misery fast. I had done this with pheasants before, often finding them burrowed deep in a soybean field after I shot them, with their eyes burning in fear and neck feathers raised in alarm. I would reach down and put my hand on their quivering back muscles and twist hard, breaking their necks. There would be one burst of flapping wings and then everything went limp. Once that border between life and death was crossed, the birds would serve some greater purpose when Mom cooked them for supper. This time, when I went from being the animal hunter to seeing the prey up close, it pulled me sharply back into being human.

Standing in knee-deep water with my father holding this wood duck, I felt a different kind of emotion, and it was not about hunter and prey. Those arguments seemed petty and pointless now. One minute the duck was flying free in the universe, angling its wings to adjust to the breeze in a marsh that smelled of algae and roots, and in the next minute it was nearly dead. Seeing it in my father's hands, the world bit into me, and the puzzling family life that I had navigated up to that point was distilled and channeled into that lone duck.

Its long, slender wings protruded at odd angles and its orange feet were splayed in gruesome directions opposite the will of its bones. I folded back one of its mangled wings and tucked it underneath its body. Delicate yellow feathers at the base of its beak were soaked in blood and blended into iridescent violet feathers smoothing its plump but tiny cheeks. Touching its feathers and feeling the warmth of its soon-to-be corpse, I had a sudden awareness of my sister Lucy. When she had a seizure, her legs splayed out, her arms flailed, her mouth frothed, and her teeth gnashed. No matter how much I tried to avoid it, my mind connected the dots between Lucy and this helpless little duck that I had mortally wounded. I tried to harden my heart and snap its neck, but I could not make myself accountable for its death. I had a sharp and wordless comprehension of Lucy's mortality: *Yes, she will die someday.*

"I can't do it," I said to my father. "You have to do it for me this time."

The duck's neck wobbled with the effort of keeping its head in the air; my father wrapped his hands around it and twisted. I heard the compound break. He released his grip and the head fell limp, but its beady black eyes were still open, staring at me. I gently touched its eye with my fingertip. It didn't respond. Everything had stopped.

"Prettiest little bird I ever seen," my father said.

I smoothed its ruffled green head feathers back into their aerodynamic position. My father put his hand behind my neck and pulled me toward him sideways until my face was pressed awkwardly into the rough canvas of his hunting coat.

"Shots ain't always gonna hit clean," he said. "You tried your best and that's enough."

CHAPTER 12

THE UNBOXING

On the last leg of my road trip, I was taking a shower at a hotel in Cleveland when Caper ate the silicone face mask of my sleep apnea machine. I took him to my veterinarian on Long Island and X-rays showed the mask winding its way through his digestive system—but that turned out to be a minor concern. The vet also discovered a heart murmur and early-stage cardiomyopathy, which would limit his life span.

"His heart is too big. It's a ticking time bomb," the vet said. "Realistically, you've got one or two good years to enjoy his companionship."

When I told Mom the news she said, "You'd better take good care of him. Lord knows, the last thing we need is another death in the family."

I took Caper for a walk on the beach, trying not to think about the specter of his heart giving out on a moment's notice. We only had to travel a few feet from my patio door and down sun-bleached pine steps to reach the shoreline of Peconic Bay. Though it was a damp thirty degrees on the coast, it felt downright balmy compared to South Dakota. The air smelled briny. A seagull dropped a clam onto the rocks and flew down to eat from the broken shell. Jetties, the wooden breakwaters that reduce beach erosion, jutted out of the rocks and extended into the bay about a hundred yards before gradually disappearing into the channel. What other hazards loomed in the depths? The surf was wild and dramatic: waves reared up and pounded into the sand, sending sea foam flying in every direction. My orange plastic kayaks made hollow banging sounds as they slammed into the bulkhead where I had tethered them. They were a gift from my ex, and I wished they would sink.

My house was separated from the bay by a six-foot-high wooden bulkhead and a sand dune overgrown with native plants: beach plums, staghorn sumac, and a type of saltwater-loving cordgrass called *Spartina*. In theory, the bulkhead was there to hold the dune in place, but the sand had shifted, causing the posts to lean sideways. I couldn't swim and had an irrational fear that the bulkhead would collapse at night while I slept, sweeping me out to sea on my mattress. Maritime deathtraps notwithstanding, maybe my luck was changing. After several minutes of straining, Caper finally expelled my sleep apnea mask on the beach. We were both relieved to be home.

On my first day back at the winery, the employees greeted me with variations of *so sorry*, *my condolences*, and *hope you're doing okay*. I tried to collect myself and run a meeting, business-as-usual, fearless leader. As the staff delivered their reports, I stopped hearing individual words. Their voices became muddled. Pumps clanged in the barrel cellar. The bottling line whirred with wine and corks flushing into glass. My mind drifted and everybody stared at me, waiting for a response to a question that I had not heard. I excused myself to the bathroom and splashed cold water on my face, thinking I just had to get through this day. A different thought surfaced when I saw my vacant eyes in the mirror. I didn't want to be there. I didn't care anymore. I didn't *feel* anything.

Then the skies broke. A once-in-a-lifetime polar vortex moved in. Snow fell for two weeks, burying the coast in four feet of white, breaking tree limbs, closing schools, and shutting down the New York City subway system. Peconic Bay froze over for the first time in forty years. It was possible to walk straight out of my house and cross the mile-wide channel between Marratooka Point and Robins Island on foot. The maddening sound of my kayaks thwacking against the bulkhead ceased as they were entombed in ice. The beach was dead quiet. As the snowdrifts piled up, work at the winery ground to a halt. I struggled to keep my driveway cleared of fallen oak limbs and became housebound with Caper during the coldest February on record.

One morning, as I gave Caper his heart meds, waited for my coffee to brew, and rubbed sleep out of my eyes, I had in mind a practical matter, a task, something to keep myself busy during the blizzard. This would be the morning that I unpacked my father's toolbox and took inventory of my inheritance.

I opened the lid and there was a photo of Lucy and me tucked inside. It was taken on Christmas Day in 1986. We were smiling from ear to ear, reclining on a yellow beanbag in front of the TV. Our blond bowl haircuts and our faces were mirror reflections of each other. I was holding a stuffed Fozzie Bear and she was wearing a white quilted bathrobe. She looked happier then and not so beaten down, with optimism still bright in her eyes and a redness in her cheeks. We had just come inside from horse tobogganing. I tucked the photo into the wooden frame of the taxidermied duck for safekeeping.

I pulled out my father's hammer and held it up to the light. It had a simple metal head and a carved wood handle with grease stains where my father had gripped it the same way for years. I wanted to know my father's hands, how he held the tools and how he felt or thought about the act of using them—if he thought anything about them at all. Over the course of the morning I removed each tool from the box, recorded it on a notepad, and laid it on the dining table next to the duck. I hoped that by documenting and organizing them, I could understand them.

They weren't expensive tools, but they didn't need to be. This was thick, tough, battle-scarred stuff meant to be held in sweaty, greasy hands. They were given life by how my father handled and carried them over the years. Some tools were asymmetric, as with the tiny crowbar that had a notched end to remove staples from fence posts. Some tools were studies in fluid movement, my fingers gliding along the smooth surfaces of shaped wood and forged iron. There was the surprising weight of a chisel, the cold metal surface of a hand plane, and the patina of bronze calipers. Others had textured surfaces, like my father's braided rodeo rope, stiff and sinewy like steel cable. There were no colors besides gunmetal and brown, save for the occasional splash of yellow on a shop pencil or the red pocket-sized tape measure from the Faith Livestock Auction. They were scratched and worn and beat-up—hallmarks of tools that were not only well used but also applied to tasks for which they were not designed. Why walk all the way across the barn for a paint can opener when the screwdriver is right there in your pocket?

The toolbox framed things in space in the way that plant taxonomists categorize botanical species according to leaf shape, flower color, and fruiting habit. There were microcollections within the whole collection: tools for clamping, crimping, and binding; tools for nailing, roping, fastening, and knotting; tools for baling and fencing; tools for fairing and leveling; tools for cutting, gouging, shaving, and carving. They covered most of the expertise that one could expect in a comprehensive woodworking or metalsmithing toolbox, with the added element of equine hoof care—farrier work. Almost everything served a dual purpose between craftsmanship and animal husbandry and had been caught in the crossfire of everyday ranch life. Growing up I saw pliers stepped on by horses, wrenches kicked under tractor tires, and hand saws squished into mud by cows. Many of my father's tools were so indelibly etched in my mind from childhood that I could never disassociate the function of them from the place and time he used them.

My father's toolbox went with him wherever work had to be done. It lived for a time on the low bench of his tack room with the horse saddlery, where light streamed in through cracked walls and cast shadows on the straw-covered floor. It was sometimes splayed open on the bench in his welding shop, surrounded by curlicue steel shavings from his drill press. When fences needed to be tightened after snowdrifts sagged the barbed wire, the toolbox sat in the back of Old Yeller, occupying space between heavy-duty towing chains and a two-hundred-gallon backup gas tank. Like the black leather medicine bag carried by the veterinarian who vaccinated our cattle, my father's toolbox was crucial to the performance of cowboy life.

Though cowboy life was never easy, my father seemed to care most for those precious few hours of solitude in which he created things with his hands. When he was doing that, he appeared focused, content, and at peace. Everything he learned about craft, he taught himself at a careful pace. He was rarely in a hurry, working with a severe kind of serenity, doing whatever handwork was required in that moment to keep the ranch running.

Other times, there was nothing serene or satisfying about his work—like when I helped him build a barbed wire fence across a steep ditch. I was sitting at the top of the ditch in the truck listening to the radio. He barked some kind of order at me—what he said, in fact, was "Bring me the stake," which he would drive into the ground with a hammer to mark the position of a new fence post. Over the truck's engine and radio, though, I thought he said, "Release the brake." I nodded back at him, then pulled the emergency brake lever. Old Yeller rolled backward, slowly at first, then gained momentum as it barreled down the ditch, tore through the fence my father had just built, and crashed into a cattle feed trough. Dogs, cats, chickens, and the odd goat scattered across the gravel barnyard in my wake. The whole scene unfolded in a matter of seconds while I gripped the steering wheel, frozen in panic. When I got out of the truck, I saw that the barbed wire had snapped and hooked my father in the cheek, causing a gash that dripped blood down his face.

He hollered at me, "When you gonna learn how to listen?"

I also learned things about myself from his tools. One day out in the pasture, right after we turned the breeding bulls loose in a herd of heifers, we were fixing the trailer hitch on Old Yeller. Out of nowhere, a temperamental Charolais bull charged at us. I darted behind the trailer and thought I was safe, but the bull rammed the side of the trailer, causing the door to slam shut on my right hand and pop my fingernails clean off. In a rare moment when my father was caught unprepared, he did not have his bullwhip, rope, or gun within reach. So he grabbed the first thing he saw. He threw tools at the bull in rapid-fire succession. He emptied his toolbox of hammers, wrenches, and pliers to fend off the bull until he could reach his electric cattle prod and chase it away for good. I ran back to the house crying, holding my mangled fingers while blood streamed down my wrist and forearm. In between screaming obscenities at the bull my father said something else to me: "You ain't never gonna be man enough."

Those words would haunt me. I would hear their echo in his voice, in the squish of hunting waders stepping into a marsh, in the metallic clinking of his wrenches while he fixed the grain combine. I would hear those words every morning when I walked to the one-room schoolhouse and watered the ponderosa pine. I would hear them when I was promoted to CEO, came out of the closet, got married and divorced, and graduated twice from Cornell University, with a master's and a doctorate, knowing my father was not present for any of it. Long after he came home from Vietnam and started fighting a different war against cancer, I would always remember that *I ain't never gonna be man enough*.

WHILE SNOW BLEW across the bay outside, I finished unpacking the toolbox and realized there was nothing else. This was all I had left of my father. The tools were proof of his life, but not much more. If this ragtag assortment had been on the shelf of a hardware store or antiques shop, I might have walked by and paid no attention. I wiped down each tool before storing it away, back inside the toolbox. There was a repetition about this act that I found oddly satisfying. Cataloguing and cleaning the tools helped me understand the scope of my inheritance, but I understood very little of the substance of it or what I was supposed to do with it.

My father could build, fix, or make anything, and for the last decade I had concerned myself with making wine. How pruning shears got handled, how grapes got pressed and fermented, how bottles got corked and uncorked, how glasses got polished—these were not just casual concerns for me. They formed the backbone of my career. I began to think that maybe my father's tools were a way back into appreciating the career choices I had made for myself, too. I did not just have to admire his tools. I could also use them to build something—perhaps a wine barrel or a wine storage rack. I would have to give it some more thought, and my best thinking happened when I was outside walking Caper.

I layered up in warm clothes, slipped Caper into his doggie vest, and put the small tape measure from the Faith Livestock Auction in the pocket of my winter parka. I wanted to feel what it was like to carry one of my father's tools with me. *Carry* would not be the right word for having a small tape measure in my pocket. That was too purposeful. It was so light and small that it shifted around my cavernous parka and mixed with my keys, wallet, phone, ChapStick, and loose coins. Such was the mindless muscle memory of everyday life and the things I carried with me. I could see how easy it would be to forget the tape measure was there—much like I had completely forgotten about the small piece of petrified wood living inside my pocket. I had picked it up when Caper and I visited the World's Largest Petrified Wood Park and Noah's Ark exhibit, and it remained there for both cross-country trips. I would intermittently put things in my pocket and take them out, not even thinking of the hard, glossy presence of the petrified wood.

I strapped on my secondhand snowshoes that I had not used in years. The big webbed feet transformed me into a hybrid creature: half man, half giant rabbit. When I opened the patio door and stepped into the white, Caper bounded past me through snow that was as deep as he was tall. I snowshoed down the bulkhead that was buried in drifts, over top of the mummified kayaks, and onto the flat, wide surface of the bay. Mounds of snow weighed down the heavy limbs of scrub pines. Their brown cones hung like white-capped eggs. With such deep snow, the trees were the only way I could tell where the land ended and the water began. I walked a few hundred yards onto the frozen bay, heading toward Robins Island, when I stopped. The wind whipped so fiercely that I feared I might become disoriented in a whiteout, so I turned around and walked along the shoreline close enough to see my house in the distance. I kept my hands in my pockets, rolling the tiny tape measure and petrified wood through my fingers, and I realized how much I cared that they had survived the journey. They were not so much inanimate objects as extensions of who I was and symbols of the past that shaped me.

Now that I had seen his tools up close for the first time since childhood and had finally hauled them across the country and set them down in my house, I found myself relieved that my father had enough of a sense of history to save them. Instead of allowing them to be lost among piles of junk, he and Mom had lifted them into that fragile space where artifacts and personal memory intertwined. While I stomped across the frozen bay in my snowshoes, I began to think of the tools as keepers of family stories that only I could unravel. While I lacked the ability to communicate with my father, I did have these tools, the strongest link to the parts of myself that I had long since forgotten.

Back at the house, I tucked the tape measure and petrified wood safely inside the toolbox. The tools sat there, staring back, silently. The mere existence of these priceless family heirlooms provided stark contrast to the mass-produced and relatively meaningless tchotchkes, furnishings, and art cluttering my house. It was funny, I thought while toweling wet snow off of Caper, we had just driven across America and back—*twice*—and lived just fine out of a suitcase, without any real hardships. I had also lived a simpler life for a few months after my divorce, when I moved my furniture into storage and got a hotel room. Why wasn't I embracing that lifestyle permanently? Most of my housewares either reminded me of my ex or were here to give the place a veneer of style. Did I really need any of it?

The next morning, I shoveled my driveway and ventured out to rent a U-Haul moving truck. I walked around my house room by room, assessing everything I had accumulated during my lifetime. Of each object, I asked myself: Do I use it every day? Is it necessary to live a healthy life? Does it have extreme sentimental value? If all three answers were no, I carried the item out of the house and put it in the moving truck. Focusing on this task over the next few days, I was shocked by how many times I answered "no." Eventually, I had emptied my entire house and filled the moving truck. My next-door neighbor, John—a six-foot-four college lacrosse player—helped me carry out the furniture and mattresses. I even loaded up the orange plastic kayaks, which may have been the most satisfying part of the process.

In the end, I stripped down my house to the bare walls and floors. The only things of sentimental value I saved were my father's toolbox, the taxidermied duck, and some old family photo albums. I also kept two suitcases of clothes, shoes, and coats; a sleeping bag and pillow; a puppy bed and bowl for Caper; one folding chair; and some basic kitchen and bathroom necessities. Without the materialistic clutter of my old life weighing me down, I had in my possession now just enough things to be able to sleep, cook a simple meal, take a shower, and look presentable at work—nothing more. The sum of everything I saved could fit comfortably inside my car. If the end of the world came, I could always escape down the highway.

I drove the U-Haul truck to the town landfill and threw everything away. In one fell swoop, I purged most of my life's possessions. I didn't want any of it. I was riding a high all the way home. I sat on the floor of my now-empty house with Caper in my lap. It was mid-February. Outside my window, the frozen bay was fringed with silver and icy blue. Everything was heavy with snow. Black dots, wintering ducks, flew past on a low-profile cruise to find open water on the sea. I felt quiet inside, quieter than at any time since my father died. It was a deep hush and yet, somehow, still restless and on edge. Around me sat a few humble boxes and suitcases containing the only things I had left in life—the things that mattered. The walls, stripped of all adornment, were a blank canvas.

I did still have one earthly possession that could be considered a décor piece, to add a splash of color and zhoosh up my otherwise bleak surroundings. I fished out a hammer and nail from the toolbox and hung the taxidermied duck in the living room. The nail I used was too thin and tore out of the Sheetrock, so the duck slid down the wall and clanged into the metal baseboard heating unit. I used my hands every day for things like opening pickle jars and shaving my head—how hard could it be to hang up a stuffed duck? There were small storage cans in the toolbox that once held Skoal chewing tobacco and Calumet Baking Powder, and now held a variety of metal fasteners. I dumped out one of the cans and chose the four biggest nails, then drove them through the corners of the duck's wood frame. I stepped back to take a look. It hung crooked, but that would have to do for now.

The photo of Lucy and me from Christmas in 1986 was still tucked into the wood frame, where I had put it for safekeeping. These two incredibly different people would always be linked in my heart: the sweet girl and the gruff cowboy. Standing eye to eye with the duck, I let my mind wander back to that cattail slough at Lewis and Clark Lake, to the day I shot it, to the magic of seeing a boat transformed and disguised as a hunting blind hidden in the marsh. It was a vision that remained frozen in time, just another casualty of my estrangement from my father. But the thought of that beat-up, rusty old aluminum canoe floating alone on a muddy lake choked off by cattails broke me open.

I believed I had been hauling only a toolbox and a wood duck across the country, but now they seemed like something more—a sign. Not only of what I could do, but of what I had to do. I had to honor the memories of my father and of Lucy. The vision was crystal clear, as if it were floating right in front of me. I had made up my mind to build a canoe.

CHAPTER 13

FEWER AND BETTER

The next morning, I didn't want to get out of my sleeping bag. The manic high had dissipated and was replaced by the worry that I had made a huge mistake—and that mistake was *not* the act of throwing away my life's possessions, but the decision to build a canoe without the faintest idea of what that entailed.

Forcing myself to get dressed and go to work was easy compared to what I faced once I arrived. Calls to the constantly ringing phone at my desk were forwarded to a recording that said I could be reached only by email from now on. I gave clipped, curt answers in a monotone voice during a budget meeting with the winery owner.

"Helluva couple months you've had. Let me know if there's anything I can do," Michael said, studying my face closely. "Are you tired?"

As bad as I felt, there was nothing he could do for me, and from his face, I realized he knew it, too. Over and over, I relived the moment where I said goodbye to my father and drove away. What had happened could not be changed, but that did not stop me from wanting to rewind back to that brightly lit hospital room and make different decisions.

Michael had to say my name to get my attention.

"Trent?"

"Oh, sorry. I'll be fine."

"Try to get some rest, okay?"

All around me the life of the winery went on. Every day, Rolex-wearing, Mercedes-driving, society-page men and women dropped by for wine tastings. They laughed and clinked glasses around the fireplace mezzanine. Though the winery was a cavernous nineteenth-century potato barn that reflected the rustic North Fork aesthetic from the outside, on the inside it was decorated like a minimalist SoHo loft, with fresh white peonies and contemporary art books fanned just so on a table. Every surface was white or black, and it seemed like it was always daytime with the sterile glow of lamplight against white exposed beams. I couldn't shake the feeling that even though I ran the place, I didn't belong here.

I had shared an office for years with the director of operations, Peter. We were about the same age and, by working in proximity alone, he knew more about me than anyone else—or, at least more than anyone who was still alive and part of my everyday life. One afternoon, Peter noticed that I was spending a lot of time staring at my computer screen perusing boat websites, wearing sunglasses so people could not see that I had been crying. He stepped closer to my desk and for an uncomfortable moment I thought he might lean in and hug me. Instead he folded his arms across his chest.

"So, is there anything you want to talk about?" he said at last. "You haven't been yourself since you got back from South Dakota."

I quickly closed the browser window displaying large-scale canoe construction blueprints.

"Oh, I'm fine—I mean, I *will* be fine. Maybe. A little rough right now," I said, taking off my sunglasses and rubbing my eyes.

"You know my dad died of cancer, too," he said, staring at the floor. "So, if you ever want to talk about any—"

"I'm building a boat," I blurted without thinking, interrupting him. Peter looked at me, skepticism flashing in his eyes before he turned and sat down at his desk.

"You're full of shit," he said.

"What?" I said, taken aback. "No, I'm serious! I got my father's tools and I'm really doing this."

"I don't believe you. Have you thought about something more, you know, *appropriate*?"

"Like what?"

"How about macramé?"

"Are you implying that I'm too gay to build a boat?"

"It isn't that; it's more that you've never been boating or built anything, as far as I know."

"I have too gone boating!" I said indignantly, though Peter was right, of course. Despite all the things I had done that struck me as vaguely related to boatbuilding, I had never actually used a saw to cut wood into pieces and then reassembled them into the shape of something useful.

"And where do you plan to build this thing, exactly?"

"My house, *obviously*—I'm right on the water!"

"But you don't have a garage."

"I'll build it in my living room. I already cleared out my furniture."

"Wow, you really have lost your mind. What kind of boat?"

"A canoe—the oldie-timey kind, made of wood."

Peter explained that canoes were more of a freshwater vessel, and he hadn't seen many paddled around Long Island. He reassured me that he was not judging me for trying to do something with my father's tools, he just thought building a canoe was an outlandish idea. He suggested that I buy a small skiff or dory to try out first, to see how I liked boating before diving headfirst into a massive project I knew nothing about. I pinched the bridge of my nose and sat for a moment.

"Just because I'm inexperienced, doesn't mean I can't try," I said.

That weekend, people stopped by to see me: my lacrosse-playing neighbor, John, who asked if I needed any more help moving furniture; people from work; and two friends from my college days in Iowa who passed through on their way to Boston. The slightly punked-out heiress neighbor in the house to the west—a woman in her seventies named Babi—showed up wearing a full-length mink coat, pink fur earmuffs, and black eyeliner and carrying a plate of homemade Italian cannoli.

"It's been so quiet over here, I just wanted to make sure you're okay," she said, peering around me into my dark, empty house. "I only know you're alive because Caper keeps shitting on my lawn."

If they dragged on too long, I broke up these visits by saying that I was tired and had work to do, but I didn't let anyone step foot inside my house. I would stand quietly with one shoulder leaning against the doorjamb and Caper peering between my legs. If they didn't take the hint, I would thank them politely for coming, but in such a way that indicated their time was up and I was not interested in small talk—except when Allen came.

Allen and I had the kind of friendship where, even if we only saw each other once a year, we would pick up at full speed wherever we had left off. He was a semifamous cable television producer and style guru who had lived in the same studio in Williamsburg for years even though he could have afforded much more. If something had earned a place on Allen's body or in his apartment, it mattered—like the midcentury lounge chair with hand-stitched avocado green cushion and carved black walnut armrests in his living room. We had been companions in adversity before: Allen's father also died from cancer. When he arrived at my house, I invited him in, nervous about his reaction. He hadn't seen my new place yet, so I led him through all the rooms and explained my plan to build a canoe. I stopped at the sliding doors overlooking the bay.

"Great view, isn't it?"

Allen was not looking at the water. He did a full three-sixty and walked back through the house, opening and closing doors, then returned to me. His eyes were wide as he pushed his tiny, perfectly round, wire-framed glasses up the bridge of his nose.

"Where is everything?"

I hesitated in responding, trying to think of something that might sound remotely sane or rational.

"Um, I cleaned house a little when I got back from the funeral."

"Not your typical housecleaning—there's nothing left!" he said, pointing to my sleeping bag on the floor. "What happened to your furniture?"

I shrugged, at this point resigned to the fact that living as a recluse would not make sense to anyone, no matter how I tried to explain it.

"I got rid of it, and now there's room to build a boat."

Allen's expression changed from shock to one of compassion. "It was the right thing to do at the right time," he said. "I know things with your father were complicated."

His skills as a producer crackled to life and he challenged me to flip the script, to try thinking of myself as a craftsman and boatbuilder first, a winery executive second.

"I know you haven't started it yet, but I want you to imagine yourself doing it first," he said, sounding like a new age self-help guru. "Get comfortable with the vision of it in your mind."

When I expressed doubt, Allen took me shopping and offered to pick out a few things to help me inch my way toward a different mind set. We went to see his friend in Williamsburg, a Korean tailor who worked out of a tiny space with a sewing machine surrounded by bolts of fabric. I took off my starched baby blue business shirt and khakis and stepped onto the fitting platform in my underwear. They set to work outfitting me in a red plaid flannel shirt and camel-colored cargo pants, which had important-looking pockets tailored to hold certain things like measuring tapes and screwdrivers, and a loop on the leg for a hammer. The pants felt stiff and itchy against my skin.

They replaced my black loafers with thick-soled, brown leather boots laced on hooked eyelets above my ankles. To finish the outfit, Allen chose an army green baseball cap that I wore with a Ticonderoga No. 2 pencil tucked on top of my ear.

I looked at myself in the mirror and immediately felt more connected to my past, resembling the roughneck men I grew up around in South Dakota, who wore practical clothing suited to their work. On the other hand, I was not in South Dakota; I was in a bespoke hipster fashion boutique in Brooklyn. In that context, I felt silly, embarrassed, like I had dressed up as a lumberjack for Halloween. Something didn't feel right about prematurely stepping into a craftsman's identity.

"I haven't even cut one piece of wood yet. I feel like a fraud wearing this."

"You'll grow into it," Allen said.

As kids, Lucy and I would sometimes dress up in Mom's clothes. There were old photos around from the time I wore Mom's blue floral nightgown and a bonnet, with clip-on earrings and a beaded necklace from her dresser. At the time, in the middle of nowhere in the 1980s—almost entirely cut off from American culture before the invention of digital cameras—I was unaware of my own image. I did not have any understanding of my relationship to masculinity and had never heard the word *gay* spoken out loud. I had never met a person from the LGBT community that I was aware of, not that I knew what the term *LGBT* meant anyway. The entire construct of being either too effeminate or too masculine or dressing a certain way to fit an image of myself simply did not exist in our bubble. There was nothing problematic about dressing up in Mom's clothes and parading around the house. We were innocent kids having fun, and who was going to see me anyway?

Out of necessity and purely for survival, South Dakotans seemed less inclined than East Coast types to draw a firm line between traditional masculine and feminine roles. Mom and the ladies from church knew how to drive tractors, bale hay, and pierce rubber identification tags through calves' ears. My father would never have scoffed at Mom's capability. She was the one who balanced the checkbook and didn't need to ask permission to buy anything. My great-grandmother Christine arrived in America from Ukraine with two oxen and a bag of wheat seeds, having no qualms about hunting and killing wild game to survive. The circumstances of our family had a way of leveling out the perceived differences between men and women.

I did see a lot of cowboys dressed in tight Wrangler jeans and leather chaps, though—and they were undeniably masculine. Their wide-brimmed hats and jingling spurs were part of the pageantry surrounding macho cowboy culture. In a practical sense, every article of clothing served a purpose: the hat shielded them from the sun; spurs controlled the horse; and chaps prevented rope burns. Removed from their functional element, though, the outfits looked odd. I was on a business trip once and saw a cowboy waiting for a flight in Chicago's O'Hare airport. He might have been the only person within five states wearing a cowboy hat and boots, but he didn't seem to care. He projected all the old-fashioned qualities of cowboys that I could not help but admire. Despite having distanced myself from that world, I still felt drawn to his air of wildness, strength, and independence. The men I knew growing up, and the men I was attracted to as an adult, were often poster images for the Tough Guy archetype in the popular imagination: Davy Crockett, Paul Bunyan, and the Marlboro Man.

I got my first taste of living up to a masculine image when my father put me in my horse's saddle for the first time. I was terrified of this beast underneath me and told my father I didn't know how to ride.

"I'm not a cowboy," I said, pleading with him to let me climb down to safety.

My father took a brand-new, kid-sized Stetson hat out of a box and jammed it on my head until the tops of my ears bent over.

"Ya *look* like a cowboy, but that's a title you gotta earn," he said, and then he slapped Socks on the haunches, and I hung on for dear life.

STANDING IN THE Brooklyn dressing room with Allen, I needed convincing that I could grow into the role I had just cast for myself when I decided to build a boat.

"It's okay for this to feel weird, but the purging is healthy, and these new clothes will serve you well as a boatbuilder," Allen said, tugging firmly on the hammer loop of the cargo pants. "I always thought you needed fewer and better things in your life."

CHAPTER 14

EDUCATED

At least one hundred eighty-two different kinds of boats exist. Most are foreign to me, and some have exotic-sounding proper names like brigantine and gundalow. The skipjack is both a boat and a species of fish. The pirogue is a Native fishing boat in the Caribbean, but quite different from a pierogi, the Polish dumpling. I discovered these new bits of knowledge thanks to Peter, whose tough questions about my desire to build a canoe triggered a search that led me down the maritime equivalent of a rabbit hole.

As a kid, I picked up a few scraps of knowledge from my father, if only because I was standing around and free to help. He had a craftsman's instinct that could not be taught in school and no amount of money could buy. I learned that the actual dimensions of a two-by-four piece of lumber are one and a half by three and a half inches. I learned how to cut ceramic tile with a diamond-blade wet saw. I learned how to dig and set post holes in concrete, how to cut in a clean edge of paint where the ceiling meets the wall, and how to make a wobbly table level with a packet of sugar. All of that, but my father never taught me a single thing about boats.

Like a good academic, I bought dozens of books about woodcraft, boatbuilding, and naval engineering, which spilled over my kitchen countertop in teetering piles. I returned to the habits I cultivated when I was in grad school, but it was not Cornell University that prepared me for study and resourcefulness. Those instincts originated at the Athboy One-Room Schoolhouse.

MY CHILDHOOD SCHOOL was a pine saltbox that had been on the prairie since the horse-and-wagon days in the late 1800s. It was a snug classroom that held ten desks, a stage, a chalkboard, a piano, and a TV on which we watched instructional VHS tapes of science experiments, like how to make an exploding volcano using baking soda and vinegar. Kids were not allowed to enter the basement, ever since the teacher discovered black widow spiders hanging from the ceiling and a garter snake hibernation den in the walls. The entire school, which housed kindergarten through eighth grade, had eight enrolled students, including me. I was the only one in my grade. The teacher's name was Fanchon Lien, and during the school year she lived in a mobile home next to the school—the only signs of human life that interrupted the flat horizon.

Every day, Mrs. Lien handed me an outdated textbook and told me to work on an assignment, which I completed lickety-split. Learning was almost entirely self-directed as the teacher allotted only one hour per day for each grade level. The rest of the day was unstructured, and that allowed me to study complex things that I did not quickly understand. I finished third-grade math months ahead of schedule and continued right into the fourth- and fifth-grade textbooks. If I read the advanced material over and over, no matter how much it seemed like a foreign language, eventually the concepts clicked. When I became bored with reading or felt that my brain couldn't absorb anything else, I went outside to roam.

One of those meandering afternoons—the day after a thunderstorm had pummeled the region with golf ball–sized hail—I saw the strangest sight of my young life. There were fish impaled on the barbed wire fences, four feet off the ground, and at least twenty miles from the nearest body of water. I touched one of them, a shimmering perch with green stripes, not believing it could possibly be real. It was dead, but still slimy, starting to dry in the sun. I burst through the door of the schoolhouse, breathless with excitement, holding a dead perch. Mrs. Lien interrupted the class to talk about my find. The most likely explanation was that the thunderstorm had rolled over the Rockies with powerful swirling winds capable of carrying the fish great distances. I was astounded. Those fish could have come from water in states to the west, from exotic places that I had only seen on maps: Denver, Bozeman, or Jackson Hole. Another plausible theory was that a loggerhead shrike, nicknamed the butcherbird, had impaled the fish there to eat them. Regardless of how it happened, from that day onward, I was captivated by the wonders of nature. If books contained answers, the outdoors held mysteries.

I didn't expect my parents to be involved in my education, either. They knew they didn't need to ask whether I had done my homework; I always had. I tried to stay quiet and require nothing from anyone. They called me a self-starter, like it was a natural gift that I had inherited from them. Instead, that was how I developed to fit our circumstances. If I wanted to do something I got up and did it, knowing not to expect anyone to coddle me with instructions or praise me with a gold star for effort. One time I left school early to move our steers from the corrals into the trucks that would deliver them to the Faith Livestock Auction. When I finished, my grandfather gave me a quiet pat on the shoulder and covertly pressed a crisp two-dollar bill into my hand. When I showed my father the money and proudly explained what I had done to earn it, he said, "Don't expect me to throw you a parade."

TWO WEEKS OF February went by, a mad expanse of days unfolding in the darkness of winter. I answered the question of how to build a canoe by poring over technical blueprints that I did not understand and reading books that were far above my beginner-level skill set. Even the word *beginner* might have been too generous, considering I had no apparent skills in anything that could be useful for building a canoe. I rushed home after work each day to sit with the growing stacks of books around me, each one dog-eared and marked with yellow Post-it Notes. I made headway on the books and reread passages fluidly that hadn't made sense when I first read them two weeks before. Unexpected lightbulbs clicked on in my head—like when it dawned on me that you paddle a canoe with a single, free-held paddle while facing the direction the canoe is travelling, as opposed to rowing while facing backward and using two oars that are attached to the boat. I shook my head in disbelief that I had made it this far in life without knowing the difference between an oar and a paddle. I had to forgive myself for not knowing what I could not have known before.

There was a brief time during my ravenous reading phase when I flirted with the possibility of building a different type of boat. I admired the elegance of sailboats, but the whole wind propulsion thing added an impossible layer of complexity. I reviled any boat powered by a motor, and for years had bristled at the noise pollution created by yachts from the Hamptons. I considered a skiff, the traditional fishing boat of the coastal Northeast. With its flat bottom, high sides, and pointed bow, the skiff had simple lines that were relatively easy to hammer together in a few days using marine plywood and basic tools. Skiffs were built in great numbers from Long Island to Newfoundland dating back at least two hundred years. In the late 1800s, Winslow Homer painted lobstermen rowing skiffs through tumultuous seas off Cape Cod, and skiffs were also the iconic spare boats lowered from the *Pequod* whaling ship in *Moby-Dick*. I found on the internet a boat shop in Maryland that sold skiff-building kits containing precut wood and held my mouse's cursor over the order button, imagining how easy it would be: I could order a kit, assemble it in a week, put my father's toolbox in the closet, and get back to living my normal life. Something about a ready-made boat kit felt like cheating, though—like saying

I had baked a birthday cake, when really, I had just opened a box of instant mix and added water. There would be no skiff kit. However clumsy I might be, I needed to learn the entire building process to feel like I had accomplished something serious and real.

The more I read, the more confident I became about my decision to stick with a canoe. It was clear that canoes were a missing link, not just to my personal past, but also to the broader history of mankind's exploration of wilderness. Native peoples set out paddling from South America to the Caribbean thousands of years before satellite maps defined the true scope of the Caribbean islands. In the year 1350, a fleet of seven canoes departed from Tahiti carrying Polynesians into the vast South Pacific, and a month later, they became the first humans to touch land on what is now known as New Zealand. The same Polynesian canoeists were so skilled that they reached the shores of Chile at least a hundred years before Spaniards "discovered" South America. Having grown up duck hunting near the Missouri River, I learned that even the word *Missouri* means "people with wooden canoes." I was blind to them my whole life, but the ghosts of distant explorers and their wooden canoes had always been right under my nose, embedded in the lore of how people explored South Dakota, and the world, for centuries.

New York has a rich history of canoeing, too. Algonquin-speaking tribes once fished from canoes in the Harlem River, near the current location of Yankee Stadium. A young Theodore Roosevelt nursed his asthma by canoeing Adirondack lakes in the early 1880s. Around the same time, a thirteen-year-old boy named Jule Fox Marshall became the first person to paddle a canoe all the way around Long Island—a journey covering three hundred miles in eleven days.

I felt a bright hum of excitement at the spectacle of canoes gliding across the water. It was a revelation: right back to human prehistory there had been people like me, bound by an inexplicable desire to build a boat and head off into the wild. I imagined the wind in my face, the paddle in my fists, knowing that I could be, for a brief moment, an heir to the age of long-forgotten explorers.

Canoes have a peculiar ability to conjure up history: in a sense, they are immortal. Few other indigenous crafts have survived the passage of time without fundamental changes. Some canoes appeared to be longer or shorter, narrower or wider, and with a more or less rounded hull, but the essence of the design looked the same to my untrained eye. Among hundreds of known kinds of boats, there was essentially only one canoe. Mine would have a silhouette indistinguishable from its ancient predecessors, differing only in the types of building materials used. That sense of timelessness, of being connected to all the canoes that came before me, felt like something solid to which I could anchor myself during this turbulent period of my life. Before long, I understood enough about the essentials of canoes, and my reading could not take me any further. I had to go outside and discover lessons of a real and tactile nature—and it all started with the wood.

CHAPTER 15

PRESZLER WOODSHOP

Large, muscular men packed lumber onto trucks lined up at the loading bays of a warehouse surrounded by a razor-wire fence. It was the first week of March. The windowless brick building blended in with dozens more like it, and its parking lot was a field of asphalt with dried weeds between the cracks and a pile of crusty brown snow plowed into the corner. Once I managed to swerve around the trucks and nose into a parking spot, I sat in the car savoring a couple more seconds of warmth and wondering what possessed me to wake up at the crack of dawn and drive to this grim industrial neighborhood by the expressway.

I had spent the last month imagining this moment while reading my boat books, but now that it was here—now that I was a few steps away from buying the lumber I needed to build my canoe—it seemed less vivid than it had in my imaginings. I felt stiff and uncomfortable wearing the heavy leather boots and workman's pants with many pockets that Allen picked out for me.

I came to this lumberyard first thing in the morning when it was only open to the trade, in part because I was curious how the place operated and wanted to support a locally owned business, rather than a big-box chain store. I also wanted to see the good stuff, the best lumber that got snapped up as soon as it was offloaded, before the yard opened to the public at noon. There was another interest at stake, too. I always had a generalized lust for wood, which stemmed from my horticultural background, and this was an opportunity to see it up close in a way I had never experienced. It almost did not matter what woods were for sale; I knew I'd want whatever they had.

Inside, a cross section of a tree trunk shaped like a six-foot-tall cookie leaned against the wall. Old-style lettering that said Roberts Plywood Est. 1977 wrapped around the edge.

"Account number, please?" asked the woman who sat behind the front desk, wearing a headset and typing at a computer.

"I don't have one," I said, hesitating, then clearing my throat, trying to sound authoritative. "I mean, I'm here to sign up for an account to buy wood."

The woman looked past me, out the glass door through which I had entered moments before, at my car parked by the entrance.

"I don't suppose you'll be hauling much lumber in that," she said while handing me a form on a clipboard with a pen attached to a beaded chain. I started filling it out.

"I can't do the business section," I said, gesturing to the form. I smiled, but she did not smile back.

"We don't typically sell to hobbyists. You'll have to make an appointment to come back during the public hours."

I made up a business name and lied on the form: Preszler Woodshop, LLC. I wrote down the address of my house, though in truth I had never considered the logistics of receiving freight deliveries in my long, narrow driveway.

"Here you go," I said to the woman, sliding the form across the desk in her direction, though she didn't turn to me for several moments.

She poured a cup of coffee while chatting with one of the loading dock workers I had seen when I drove in. She finally turned and looked over my form.

"What kind of business is Preszler Woodshop?"

"Boatbuilding," I said, uneasily.

She checked her watch. "I'll page someone out back. Have a seat."

While waiting, I flipped through assorted magazines and do-it-yourself manuals on the coffee table with names like *Popular Woodworking*, *Fine Homebuilding*, and *The Little Book of Whittling*. None of it was all that interesting to me and I was about to look at my phone when I saw a paperback. I fished it out of the pile and read the longest title I had ever seen, printed in six different sizes and colors of type: *Building a Strip Canoe: Full-Sized Plans and Instructions for Eight Easy-to-Build, Field-Tested Canoes*, *Second Edition*, *Revised & Expanded* by Gil Gilpatrick, Master Maine Guide. Sandwiched among the many words of the title was a photograph of a wooden canoe floating on calm water. In all my research about boats, I hadn't come across this little gem, which I attributed to the fact that it had the weight and feel of an oversized instructional pamphlet rather than an actual book. It was a reprint of a DIY manual with an original publication date of 1983, and that is precisely what made it interesting. I had grown up teaching myself with ancient, secondhand textbooks, and *Building a Strip Canoe* felt familiar in a homespun way.

I read the back cover, which described the author, Gil Gilpatrick, as a man who had built more than five hundred canoes during his thirty-year career teaching high schoolers in Skowhegan, Maine. The blurb said that the author's many years of experimenting had "refined his contemporary construction method of augmenting traditional wood strips with fiberglass and epoxy." If teenagers could do it, my odds had just improved.

I flipped through the pages and looked at the dozens of color photos and illustrations of canoes in various stages of construction and use. The images were poorly lit and grainy, obviously taken before the advent of high-resolution digital photography, and not updated since. One of the photos showed a canoe floating on tranquil water with spruce trees in the background. In another shot, a woman with bleached and permed hair who was wearing a yellow rain slicker paddled a canoe with her male companion, who sported a handlebar mustache and acid-washed jeans. The smiling owners paddled, surrounded by camping gear. My eyes rested for a moment on a photo that delivered a silent gut punch: a father and his son paddling a canoe down the Allagash River. I assumed they had built the canoe together.

I weighed the book in my hands, turning it over. Something about it felt reassuring, like one of Mom's *Betty Crocker Cookbooks*: illustrated, prescriptive, and practically foolproof. The other books I had at home were too theoretical and complex to be of much practical use as I figured out how, exactly, to build a canoe.

"Trent, they're ready for you out back," said the woman at the desk, waving me through the double doors into the warehouse. I carried the book with me and walked into the warehouse, where she instructed me to wait behind the metal safety railing. Inside was the craftsman's version of Disneyland, Oz, Santa's toy shop, and Willy Wonka's chocolate factory, all rolled into one. *This*, I told myself as I inhaled the spicy aroma, *is where wood comes from*.

Men in overalls and steel-toed boots hurtled around on forklifts in an enormous fulfillment center. The aisles rose on girders into an endless chasm of wood that was accessible on foot through steel-grated stairways and catwalks, like fire escapes spreading across old New York City brownstones. The trade floor bustled with activity and no one seemed to notice my presence as they grappled with massive timbers of countless shapes and sizes. Workers unpacked mountainous pallets of lumber, scanned them into a computer system, and moved them to precise locations in the vast, 3D storage matrix. Each cubicle was labeled from floor to ceiling with the species of wood stored there, and the names progressed alphabetically around the room, from afromosia to ziricote. Hundreds of chocolate-brown boards were hoisted onto a tier of steel shelves in a cubicle labeled for black walnut. Entire tree trunks six feet in diameter were sliced lengthwise just the way they left the forest, with each slab separated by thin wedges. There was a metal floor stand holding thick slabs of maple. Another forklift rolled by, carrying more white-bark birch logs than I had ever seen in one place. There were stacks of kiln-dried Italian ash, Brazilian purpleheart, and Guatemalan cocobolo. My eyes glazed over with this extraordinary collection of lumber. Somewhere among the stacks, resting inside the millions of board feet, there must be a few humble planks that could one day become my canoe.

A man in a white collared shirt who appeared to be the boss of the place peeled shavings off a giant slab using a hand plane, revealing the bright orange and rare exotic wood underneath the dusty surface. A prominent Hamptons building contractor, recognizable from the logo covering the back of his polo shirt, leaned in to sniff the freshly planed slab. His client stood at a safe distance with her face half-concealed by designer sunglasses. I could not tell what they were saying, exactly, but I knew they were negotiating. The woman shook her head, and the salesman remeasured the dimensions of the slab. Then he tapped on a calculator and wrote the price on the wood using a piece of white chalk. The contractor looked at the woman. She shook her head. The salesman again tapped on his calculator, then crossed out the first price with a dramatic *X* and scribbled a new price beside it. The woman seemed satisfied. With a nonchalant wave of her hand, she purchased the African padauk slab that cost as much as a car. Gaining an insider's perspective on the commerce of wood auction markets—this massive bulk commodity churning dollars into the American economy—was in many ways similar to my experiences as a young boy at the Faith Livestock Auction.

WE CALLED IT the sale barn, and it consisted of acres of steel holding pens to organize cattle according to sex and dozens of breeds, weights, and ages. There were pens for yearling black Angus heifers, hulking red Limousin bulls, lanky longhorn steers, and speckled roan Hereford cows. The pens funneled into the auction arena, where the floors were covered with cedar shavings piled ankle-deep to absorb cow manure and the tobacco juice spit out by cowboys.

Cattle auctions were a normal part of our seasonal ranch life, but to outsiders it must have looked quite peculiar. On occasion, tourists would show up to take pictures and stare at us with wild eyes. They were people who lived in cities on the coasts and had never worked on a ranch, and for whom the meat in deli sandwiches seemed to magically materialize without agricultural labor or an animal's death. My father was proud that the beef cattle we raised got shipped around the world. The first time he saw the Wendy's fast food chain's "Where's the Beef?" ad campaign, he stomped his heel on the ground and said, "Where'dya think all that danged beef comes from, anyway? Right here!"

On auction day in November, the whole Preszler extended family filed into the sale barn amphitheater. Everyone was covered in a thin film of dust. The whole building smelled like a giant hand-crank pencil sharpener. Mom let me fill a small bag with cedar shavings from the floor, which I opened periodically and smelled. Neighboring ranchers tipped their hats when they passed.

"You're Leon's boy, ain'tcha?" the ranchers would say, and I would smile back timidly, hiding behind the brim of my cowboy hat. Out here, a son was forever known by his father's name.

"Aaaand we've got this season's weaned calves from the Preszler ranch," the auctioneer said into the microphone, his breath a white puff in the cold air. "Let's give'm a good sale, folks."

The auctioneer barely opened his mouth, but the sound that came out was something between a yodel and a rhythmic, rapid-fire narration of price per pound, with filler words peppered into the flow. Once I tuned into his rhythms—and it took a few minutes to acclimate to the crests and swells—the prices became discernible: "One-dollar-quarter bid, now a half, now a half, do I hear one-dollar-half? Standing one-dollar-quarter for the Preszler family in the top row, let's give'm a nice run, folks. YES! Over there, gentleman from Hormel, I have now one-dollar-three-quarters bid, one-dollar-three quarters—now TWO, now two, would you give me two?"

"It's cowboy poetry," my father said.

People in the crowd raised their hands to bid. The bidding went back and forth for quite some time. As the price inched upward, the tension in the room grew.

"SOLD! To that Oscar Mayer fella from Chicago," the auctioneer said.

Our neighbors glanced over their shoulders and tipped their hats to my father. They knew what his name was worth.

THE BEARLIKE MAN approached me from the sales floor at Roberts Plywood.

"I'm Scott Roberts, the owner, and yes, I have two first names," he said, almost crushing my hand when he shook it. "I understand you're a boatbuilder—what kind?"

I would just have to roll with this made-up Preszler Woodshop thing. "Canoes," I replied. "Well, I mean, just one canoe."

"Old-school. Don't see many of those around here. What got you into that?"

"When I was a kid, my father took me out on an aluminum duck blind."

"Ah yes, Grumman canoes—built like tanks in a factory ten miles down the road from here. How long have you been a boatbuilder?"

"Um, maybe about a month or so?" I replied, exposing my ruse.

"Oh brother," Scott said, rolling his eyes. "We don't normally spend much time with hobbyists. Tough to get people to commit to anything substantial when I'm sending full trucks to commercial contractors in the city every day."

"Look, I promise not to waste your time. I have a PhD in horticulture."

"A plant doctor? Well, in that case I'll make an exception. You'll be my only client who knows the Latin botanical names of the woods."

I followed Scott around and we talked about the pore, luster, and color of wood. I held boards in my hands and noted their different weights and grain patterns. He pointed out the ribbon stripes of tiger maple and the frothed grain of burled walnut. I admired a piece of sycamore that resembled a crocheted lace doily. We discussed the difference between ebonized wood and true ebony, which is so dense that it sinks in water.

"Sometimes, when I'm not sure what I have, I'll do a scratch 'n' sniff test," he said as we continued down the aisle past shelves loaded with cherry and chestnut. We stopped next to a stack of maroon-colored lumber with white streaks running through it. "This one fills the warehouse with its aroma after you cut it."

I leaned over and inhaled its cedar-like scent. Images from a different time and place flashed through my mind. I had encountered this wood before.

"That's true juniper, *Juniperus virginiana*," I said. "Wild junipers were just about the only trees that survived winters where I grew up. My father used the trunks as fence posts."

"I'm impressed you knew that. Clients misidentify it as cedar, so I stock it here with the cherry and chestnut," Scott said. "It'd get lost next to jatoba."

"I thought I was the only person who cared about those distinctions," I said. "Names matter."

"Cedar trunks, cedar closets, cedar in sock drawers—people ask for cedar and don't realize what they're getting is juniper. If it smells right, that's all they care about."

"Same thing happens to me at work. People ask for red wine not knowing the difference between merlot and syrah."

Scott picked up a small board from the stack and slammed it across his knee. It broke into several pieces. "Juniper is brittle because it has so many knots interrupting the grain," he said. "If you can avoid the knots, it's not so bad, but I've always preferred hardwoods."

With one smooth tensing of his shoulder muscles, Scott lifted a bigger board onto sawhorses and marked the width to cut it in half. It must have been milled from an old tree. The slab looked to be about two feet wide, eight feet long, and three inches thick. I started to ask why he adjusted the lever on top of his circular saw, but I stopped. He looked deep in concentration and I didn't want to sound like a novice by asking a basic question. He made the long cut and paused to insert a small wedge of wood on the part he had already cut. He worked with the precision of a sushi chef and, when finished, he struck the board with a mallet and it fell into two perfect pieces.

"What was that wedge for?" I asked.

"You need that on a rip cut so the blade doesn't bind."

"Oh, that sounds like an important thing to know."

"Woodworking 101, my friend."

"You think I could use this for my canoe?"

"I always recommend western red cedar for boats. It's softer, pliable, easy to work with, and comes in any length you want. This juniper is gonna be pretty short for you. At any rate, I don't stock many soft woods."

I had already made up my mind. The smell of juniper and sight of its red-orange wood had taken me back to the ranch days. I had to have it.

"I'm going to use juniper. I'll take a whole pallet."

After another hour of walking around with Scott, I also selected some black walnut and creamy white basswood for accent colors.

"This book is coming in handy already," I said to Scott while reviewing the shopping list in Gilpatrick's *Building a Strip Canoe*. "Add it to my bill."

I picked out a few other things, including construction-grade Douglas fir beams to build the foundation of the canoe. On the way home, I stopped by a hardware store to fill up my car with everything else on the building list that was not wood. What a massive list it was. The Roberts Plywood delivery truck arrived later that day and slowly, carefully backed down the narrow driveway to my house. Two strapping guys in canvas overalls and a giant man with a Viking profile and the girth of a middle linebacker unloaded the lumber into my house. It was not so empty anymore. I signed the delivery receipt and the Viking looked around.

"Nice place," he said, gazing at the bay. "You're lucky to live on the water."

That night, I ate dinner in my chair surrounded by waist-high stacks of lumber and bags and boxes of boatbuilding supplies that I had not yet removed from their packaging. They sat on either side of me, corralling me in, covering almost all of the available floor space in the living and dining rooms. Caper lounged on his bed between the lumber. I pet him and touched the top of the wood stack as if I were caressing the heads of my two children.

My crazy idea to build a canoe had progressed far enough along that I was past the point of no return. I had no choice now; I had to tell Mom what was happening. I called her and explained that things had changed since the funeral—things I sensed in my gut but didn't know how to put into words.

"I think I'm on a different path now—maybe better, but scary," I told her. "I cleared out my house to build something with Dad's tools."

"Oh good, a project! What are you making?"

I picked up *Building a Strip Canoe* and admired the cover photo of a canoe floating on the calm, blue surface of a lake.

"A wooden canoe, and it's going to be beautiful."

She went silent for a spell. "I thought you would've started with something smaller, like, say, a cutting board for my kitchen."

"I need to do this for myself, Mom."

After we hung up, I reached into one of the plastic shopping bags and pulled out a dust mask, whose packaging claimed the mask would capture ninety-nine percent of harmful dust particles. I ripped it open and held the mask up by the elastic straps, then put it around my face, as if I were a doctor at a hospital. I might have put it on backward because the filter cartridges jabbed into my cheeks. Was I supposed to saw wood wearing it like this? It seemed silly, but I didn't know. I took off the mask and tied it to the frame of the taxidermied duck, so it would dangle on the wall while I worked nearby—easy to reach should I need it.

I grabbed the fist-wide badger-hair brush from my father's toolbox and a can of chalkboard paint that I got at the hardware store. I pried it open with my father's screwdriver. It took an hour or two of concentration to cut in paint with the brush along the baseboard heater edges and where the ceiling met the walls. On the main surface, the paint spread creamy and thick, a silken gloss that dried to coat the walls, minute by minute. With the living and dining room walls now charcoal gray, I stayed up late and immersed myself in the pages of *Building a Strip Canoe.*

Gilpatrick wrote about the milestones of the building process and something called the golden chain of sequential construction. If his time-tested construction method was not executed carefully, he warned, any mistakes in the early going would be compounded on each successive step. Like when laying bricks, if your bottom row was crooked, the whole wall would be off, too. The first step was to ripsaw twenty-foot-long strips of juniper, a quarter-inch thick, and glue them onto forms to create the hull. That's where the strip canoe got its name: from the wood strips used to build it. Next was sanding the hull smooth and fusing the strips together with a watertight epoxy-fiberglass composite. The last step was to install hardwood trim pieces and brush everything with glossy marine varnish. The whole canoe would take an experienced builder two or three months to complete, working full-time, but I still had my day job at the winery. Realistically, for a complete novice like me, the whole process would probably take between six and nine months.

I had a moment of panic facing the monumental task ahead of me. It was already March and there wasn't enough time to build the canoe and paddle it this summer. There was another milestone in the nine-month time frame, though: early December. Assuming the bay did not freeze again this year, what if I finished in time to paddle it on the anniversary of my father's death? I reverse engineered the build-out process on a reasonable timeline that I considered like it was a winemaking challenge: first grow the grapes, then ferment the grapes, bottle the wine, and finally drink the wine and savor the fruits of my labor—a sequence of the seasons. I could do this.

After the paint dried, I found some broken fragments of chalk inside a tobacco tin in my father's toolbox—apparent relics from the schoolhouse. It was the middle of the night, but I had no use for time. The compulsion to start was too strong for me to resist. Between two and four in the morning I marked up the walls. In the dining room, I traced an oversized wall calendar with big squares for the months, weeks, and days. Working line by line using a piece of wood as a ruler, I plotted out a grid for the rest of the year, with a big circle around the first week of December. The next job was bigger, covering the entire living room wall. Up on a stepladder, I freehanded eight capital letters, tracing their outlines first before filling them in with whatever chalk stub was left and smearing the white dust around with my fingers. I stepped away to reveal the word. *F L O A T I N G.*

At chest height, the curves of the letter *O* resolved into a two-foot-wide frame around the hanging taxidermied duck. I had filled the wall with things that could not be walled in. The duck would be forever flying, and I hoped to someday be floating on the calm, blue waters of Peconic Bay, just steps out my back door.

After rinsing off the paintbrush I stuffed it back into the toolbox and lay down in my sleeping bag to catch some rest before work. At first I worried that I would not be able to sleep with my father stalking my dreams, calling to me from the stupor of his hospital bed. Instead, I thought of the sweet smell of wood shavings, and of *Juniperus virginiana*. Everything was different about my house now. It was the wood. The wood filled my house with wildness like it had filled the Faith Livestock Auction with its scent. I shut my eyes. It was about to begin.

CHAPTER 16

THE STRONGBACK

Canoes are not composed of anything that might seem simple to measure or cut: no straight lines, no flat surfaces, and no right angles. They're connected to nature by design, with curvilinear hulls resembling living sculptural forms in the sea, like large fish and limbless mammals. To create such elegant curves, I tried imagining myself as a sculptor, starting first with a solid block and chipping away, down lower and lower, experimenting with different shapes as I went, settling on one when I reached the perfect curve. That was a ridiculous thing to imagine, though. As I quickly learned, canoes are not sculpted. I would not start with something and whittle it down. I would have to build something up from nothing.

The range of boatbuilding choices available to me had narrowed and the path forward could be discerned. I shackled myself to the pages of Gilpatrick's book. His method was the only method now. Small instructions followed by small tasks would help me transfer the designs from those grainy 1980s photographs onto the lumber that filled my house.

A canoe gets built on top of a rigid support base, called a strongback. This all-important structure resembles a twenty-foot-long wooden stepladder lying flat on the ground. To form the side braces of the strongback, I had purchased two-by-eight-inch construction beams that were milled from Douglas fir trees from Oregon's Pacific Coast Range. When I laid them on the floor parallel to each other, they spanned the entire living and dining rooms, with a couple feet to spare on each end. I would join them together with twelve short blocks of wood attached to the beams eighteen inches apart, like rungs on a ladder. It seemed simple enough.

I picked up my father's metal triangle and the miniature red tape measure from the Faith Livestock Auction. The triangle was cold in my hands and heavier than I expected. I had played with my father's tape measure as a kid, clipping the end to a leg of the dining table and extending it across the floor, then hitting the spring release button so the tape whipped back and caused the holster to spin from the momentum.

"Yer gonna bust it, and where'd we *be* without a tape measure?" my father would say with a tap on my shoulder.

I stretched the tape down the length of a two-by-four and made a pencil mark every foot, then I used the lip on one side of the triangle as a straightedge to draw a line at each of those hash marks. The blade of my father's handsaw was flimsy, rusted, and kinked in the middle, so only half of the teeth were usable. I tried sawing with it anyway out of a sense of obligation. I gripped the two-by-four with my left hand and plunged downward with my right, holding the saw against the pencil line. The blade skittered off the board and nicked the skin between my left thumb and forefinger. Less than an hour in and I had drawn first blood.

Fortunately, the first chapter of *Building a Strip Canoe* featured a comprehensive list of power tools I would need for the build. Not one to pass up a shopping opportunity, I had dutifully purchased everything on the list—things like a table saw, jigsaw, miter saw, sawhorses, power drill, and a snap-together workbench. I removed these objects from their shrink-wrapped boxes covered with neon-colored brand-name graphics and arranged them around the house in the last remaining slivers of exposed floor space between stacks of lumber. Every box I opened and power tool I plugged in left my house cluttered with cardboard and left me pulsing with monstrous guilt. By the time I finished unboxing everything forty minutes later, I was practically morose. There, spread around the house, was a suburban DIY weekend hack shop of power tools. All of this meant my father's humble hand tools were insufficient for the task.

I had made a promise to myself to use his tools based on a naive assumption that everything required to build a twenty-foot canoe could be found inside a two-foot toolbox. I had overlooked the practical notion that specialized crafts had their own specialized tools to accomplish specialized tasks. It was clear that a cattle rancher's tools were not all going to translate into boatbuilding—different crafts, different eras, worlds apart. Nonetheless, I had to defer to Gilpatrick's knowledge.

It was time to start. After resting a two-by-four at waist height between sawhorses, I plugged in my new circular saw, lowered it onto the wood, aligned the teeth with the pencil mark, and squeezed the trigger to bring the blade into a spin. It produced a frantic, panicked, piercing sound. As the spinning blade touched the wood, its desperate wail shifted, rising to a roar that blasted through my squishy orange earplugs. The saw chewed a dark line, spraying sawdust into my face. I tried to keep the saw straight, shifting it when necessary, knowing that a brief lapse in concentration could mean a finger on the floor and blood everywhere. Somehow, the wood pinched the blade and the saw came to a shivering halt with grinding noises that signaled *wrong*.

I washed the sawdust out of my eyes and put on safety goggles, then I backed the saw out and tried again. It sputtered to life and I pushed it forward, cutting all the way through. When the first foot-long piece clanked to the floor, I was relieved. On I went, measuring, marking, and cutting a dozen pieces. The sawdust made me cough until I remembered the dust mask hanging from the taxidermied duck and strapped it on. I cut one piece too short and set it on the floor in the corner. I cut another piece with a sloppy and jagged end, so that went in the corner, too. At some point, I would have to light a beach bonfire with all the offcuts. The pieces that were the right length made a satisfying *tink* when they broke off from the main two-by-four. Sawdust settled down onto the tile floor, resting in the grout seams. My house filled with the scent of Douglas fir, bright and clean, the smell of Christmas.

The instructions to attach the crosspieces to the beams included the use of a power drill and screws, and not my father's tools. Something else was there right in front of me, though, and capable of joining wood. That's what a hammer and nails are for. The business end of my father's hammer, just below the metal head, was wrapped in many layers of duct tape. I could not tell if the tape was decorative or served some purpose like reinforcing a cracked handle, but I thought it best not to remove it and find out. A small engraving was perfectly centered on the butt of the handle: "USA HICKORY, LKP." The latter was my father's initials, for Leon Kenneth Preszler. A little burst of pride swept in and some of my tensions dissipated. He had held this hammer in his hands for horseshoeing, house construction, and everything in between, and now it fit my palm like a well-worn glove. Even though our hands had always *looked* identical in photos, I now knew whatever tools my father could grip, I could grip, too.

I stretched the tape measure down the length of the two-by-eight beams sitting parallel on the floor and marked lines every eighteen inches where the crosspieces would be attached. I pinched a three-inch nail and positioned it near the end of the first crosspiece, so that when driven down it would also pierce the beam underneath it. I tried to press the nail tip into the wood to give myself a head start, to stabilize the nail before I took a whack. I struck down. The nail blasted off, skidded across the floor, and clinked into the baseboard. I grabbed another nail—a quick salvage mission for my self-esteem. I got down on my hands and knees, closer to the wood, nail in position, *bang*, *bang*, *bang* . . . and the fourth strike glanced off the side of the beam.

The hammer came down hard on the floor and shattered the ceramic tile into cracks that radiated out from the point of impact. I sat back on my heels. *Son of a bitch.*

GROWING UP POOR in South Dakota, I collected aluminum cans for recycling money—a standard weekend activity and one of the ways I made myself useful around the house. I often tagged along with my father when he competed in summer rodeos around the Great Plains. In the shadows underneath the spectator bleachers and behind the bucking stock chutes, I would spend entire weekends hunched over, picking up empty pop and beer cans. While I worked, a line of anxious bucking broncos clanged into the chutes behind me, and the rodeo contestants carried their saddles past me in search of their assigned rides. The black plastic garbage bags that I used blew in the wind and stuck to my sweaty, dusty legs. My hands would grow brown and sticky. A few hours later, with the bucking corrals empty and the crowd gone, I would kneel in the dirt and scribble numbers with a stick, counting my haul. Sometimes I would have so many bags that the whole truck bed filled up, and we would drive to a recycling center and cash in.

It dawned on me that if I could find a way to crush the cans flat, I could squeeze more into the bags, and we could make more money. The most logical way to flatten a truckload of cans was to set them on a hard surface and smash them with a hammer. I searched for a hammer in my father's work garage, which smelled like oil and sawdust and was crammed with unidentifiable machines plugged into extension cords. He had built the garage next to the corrugated metal grain bins where he parked the tractor and grain auger. I stepped around big circular blades and an iron anvil as tall as I was and walked across the part of the dirt floor that was always soaked with blood. There was a hook and chain hanging from a pulley in the ceiling. My father would sometimes bring home big bleeding mule deer and peel off their skin with his knife here. I crouched in the fine peels of horse hooves and peered into the shadowy space under a shelf. That's where I found his toolbox. I pulled out his hammer, tucked it under my T-shirt, and scrambled back across the barnyard.

I dumped the garbage bags just outside the kitchen door, onto the redbrick patio that my father had built the previous summer. I started swinging the hammer and did not stop until every pop can was flat as a pancake. Upon returning the cans to the plastic bags, I was able to stuff in three times as many. During that process, though, I also smashed many of the bricks on the patio, reducing them to pulverized dust, and breaking the cement bonds holding the patio together. Though I knew what I had done was probably not great, I rationalized that any damage to the bricks would be offset by the good I had done for the family. I had made us some money. On the balance, I was proud of my work.

Later that day, I was peeling potatoes in the kitchen while Mom fried the pork chops that had bled through white butcher paper after thawing in the kitchen sink. She became preoccupied with Lucy, who was doodling in her coloring books and got Crayola markers all over the dining table. My father pulled into the gravel driveway in Old Yeller after spending the day in the fields harvesting wheat. I watched from the kitchen window as he got out of the truck, took off his sweat stained cowboy hat, and yawned. Underneath the windmill, he fished out a Pabst Blue Ribbon from the metal stock tank, cracked it open, took a long drink, and stared off in the distance with a weary look.

I bolted out the kitchen door and waved him over to the patio.

"I've got a surprise for you!" I said.

He sauntered over, holding his beer. He was sunburned, his eyes tired. Bits of wheat chaff clung to his face.

"I crushed all the rodeo cans so we can fit more in each bag," I said, excitedly pointing to the bags slumped over the edge of the patio. I walked toward him holding his hammer in one hand and a flattened pop can in the other. "Look, see?"

My father clenched a fist and punched me in the face.

"What in the hell didja think yer doin'?" he screamed, red-faced, the vein on his bald head popping out. He grabbed the hammer out of my hand and shook it in front of me. "You ain't never touchin' this again—and this here patio was all we got, boy. You better learn that right quick if you learn nothin' else."

My legs shook in my cowboy boots. We stood that way for a long time, facing each other, my father's barrel chest rising and falling with his angry breathing. I could not speak a word. It felt like minutes ticked by. I was completely unaware of the fact that I had peed myself, until I saw the puddle soaking into the sand between the broken bricks. I did not mind so much that my father saw me wet my pants, but I refused to let him see me cry.

Then, just as suddenly, we torqued things back to where they had been. There was three feet of distance between us and nothing to do but walk away in opposite directions. I thought it would be no news, our little secret, nothing ended or resolved, just an explosion and a fizzling out. I didn't tell Mom. I hid the wet jeans in the bottom of the laundry hamper.

That night, I sat down for family supper. The left side of my face felt hot and throbbing, but I didn't dare touch it or act like it bothered me. I played it cool. On the table, pushed off to the side to make room for our plates, lay seed catalogs and the *Farmers' Almanac*, Lucy's day-of-the-week pillboxes, little ceramic salt and pepper shakers of brightly painted pheasants, a bowl of salted peanuts in the shell, and Publishers Clearing House Sweepstakes mailings opened with hope. My father sat down without saying a word. Mom took one look at me and pointed at my father.

"Did you hit him?" It was the only time I heard her raise her voice.

"You darn betcha, and I reckon he deserved it."

"He was just trying to do something good, so we could have a little extra money."

They argued while I poked food around my plate, not eating. Worse than the purple bruise swelling up on my cheek was the betrayal I felt in my heart. While it was not uncommon for my father to spank me when I got in trouble, he had never gone this far.

"Sorry I hit you," he said. With his deadpan voice, I could not tell if he was just saying it to appease Mom, or if she had convinced him he was wrong. Even this stoic wall of a man could be temporarily swayed by her firm and insistent reasoning.

"Good," Mom interjected on my behalf. "Now that that's settled, everyone should have more potato salad. Lord knows I made enough for leftovers."

IN ALL THE years that followed, my father never laid a hand on me again. I had always supposed it was the static in his head that caused him to snap, some terrifying noise the rest of us couldn't hear, perhaps the secret tensions he held in from Vietnam. He changed under the pressure of our family situation. His spirit flickered over the years as our financial stability fluctuated, Lucy's health deteriorated, and I became more of an independent man. That day marked the beginning of a permanent shift in my own perspective: where I had once seen a rugged, heroic cowboy father, I was now more likely to see a struggling, flawed father whom I could not trust to keep his anger in check. When I inherited his tools, I was torn between honoring his legacy and doing something different for myself, to create some separation from the darkness. In a way, the canoe project offered both.

I SWEPT UP the broken shards of ceramic tile from the floor next to the strongback. With the claw side of my father's hammer, I pried the mangled nail out of the wood, then held a new nail between my thumb and forefinger. This time, seven *bangs* and the metal disappeared into the wood. The two-by-eight beam was attached to a short segment of two-by-four. The strongback looked like a giant letter *H* spreading across the house. One down, eleven to go.

I learned to start out with smaller whacks and gradually increased my swinging force to full power. Something was getting done here. Something concrete and specific that mattered to me was happening.

The nailing was often just bent nails and bruised fingers. Mostly it was hard work. A dime-sized blister developed on the pad of my thumb. The muscles in my arm burned as my tennis elbow flared up.

When I did it right, though, hammering moved my whole body and quieted my mind. By the sixth or seventh crosspiece, I got the hang of it. With each downward stroke, I could only think of one word: *yes*. One by one, the steel shafts disappeared into the Douglas fir and only their heads remained visible: *bang*, *yes*, *bang*, *yes*.

Of all the small remnants that might have survived beyond my father's life, none could have been more useful than this magnificent, greasy, hickory-handled, and duct-taped hammer. Every swing, every nail, and every fragment of energy I thrust into the job fortified my sense of purpose and made me feel a little less broken, a little more capable, and a lot more forgiving.

So, by nightfall, two parallel beams joined by twelve crosspieces sat on the floor of my house. Once I stopped hammering, Caper emerged from the back den to see what all the ruckus was about. I found *Building a Strip Canoe* lying on the floor covered by sawdust. I picked up the book and placed it on top of the strongback, opened to that page. A little rotation, and the outline of my strongback matched the one in the photo. I felt a sudden levitating sense of having completed the first step.

Given the wide range of objects in this world that I might logically have constructed using three pieces of wood and some nails, I never would have predicted it would be a canoe strongback. However rudimentary this structure was, it gave me assurance, for the first time, that I could build something with my own two hands. I could cut big pieces of wood into smaller pieces and reassemble them into useful shapes. Was that not the essence of craft?

I spent an hour sweeping, organizing, putting things back in the toolbox, and making neat stacks of the offcuts. I took a photo to remember the moment. Until just days ago I was terrified of the whole project, but the fog had lifted for a few hours, and I forgot my fears. The strongback had led me to something I had *not* forgotten: why my father's hammer mattered to me.

CHAPTER 17

TRUE TO FORM

It was almost April. The days were getting longer, and it became a regular part of my evenings after work to admire the strongback. This little island of woodcraft cheered me up. Nailing two-by-fours together at right angles was a silly thing to get excited about, though. This small triumph brought with it a whole new set of worries—namely, how to build this canoe beyond the elementary stage into the curvilinear realm.

There's a method to making the multidimensional curves of a canoe, as it turned out. The full length would measure twenty feet, with the hull, or body plan, divided into twelve equal sections, called stations. If the canoe were imagined as a loaf of bread, then each of the twelve stations would be a slice. Those cross sections are called forms, which are made by tracing design patterns onto plywood and cutting them out with a jigsaw. Once the forms were cut, I would attach them to the strongback. Only then, with the full architectural base set in place, could I start bending long, thin strips of wood across the forms to create the curvilinear hull.

Bright spring sunlight beamed in through the patio doors and across the kitchen floor as I unfolded blueprints that were tucked inside the back cover of *Building a Strip Canoe*. Caper, long faced and floppy eared, sat nearby, watching me. The sheets were preprinted with outlines of the forms—Gilpatrick's foolproof method for tracing accurate shapes onto wood. The widest forms would be installed in the center part of the canoe (amidships), and the narrowest forms—tapered, six-inch-wide ovals—would be installed toward the ends of the canoe in the front (bow) and the rear (stern).

I taped the first pattern to a sheet of plywood. With a few strokes of a pencil, I was off, tracing around the curves. I pressed so hard, and the plywood was so rough, that I had to stop to sharpen the pencil several times. This tracing exercise was not the artistic immersion in boatbuilding that I expected. As a novice, though, I clung to the certainty of mechanical tracing. There was comfort in being bound to plans, calculations, and measurements that were predetermined by someone who knew what they were doing. Any kind of flippant alteration to the hull's shape would not be true to its history or function and could result in disaster. Each pattern I traced anchored my canoe to boats that builders before me had already proven could float. From the indigenous peoples who had first invented canoes independently of one another on different sides of the globe, to the North American explorers who refined their designs, to the modern boatbuilders who constructed canoes with different materials—canoeists had been using some version of these patterns for centuries.

I started cutting around the first penciled outline with a handheld electric jigsaw. The term for using a jigsaw is *to cope*, and so *coping* is cutting curved shapes in wood. As the blade thumped up and down through the plywood, my hands, elbows, and shoulders vibrated. Wood that was once whole turned into something divided. It was harder than I expected to guide the jigsaw with steady hands at an even pace, and as a result, my coping line bobbed and weaved around the curved edge. Three jigsaw blades snapped, and I ran to the hardware store to buy replacements. Sometimes the blade chewed a jagged bit on the outer edge, corrupting the smoothness of the curve—a mistake I winced at, but moved past. On more occasions than I cared to admit, the saw got away from me, the tracing line blurred under the sawdust, too much wood was taken off, and there was a gouge in a part that was supposed to be flat.

When my coping went well, my movements were slow and accurate as I traced the lines. My vision locked in on the wood disappearing against the blade. I focused on my breath flowing in and out of the dust mask. In those moments, I lost track of time and place, and I learned to trust my abilities a little more. Rounded forms popped out of the plywood like the shapes Mom made pressing her cookie cutters into rolled dough. Bits and scraps fell away and soon enough I had created a jumble of curved shapes stacked on the kitchen counter. I numbered them one through twelve in the sequence they would be attached to the strongback, from bow to stern. I wiped sweat off my face and tried to brush the sawdust off my arms. It clung to my skin like sand.

Attaching the forms to the strongback seemed like a simple, straightforward exercise: just screw them to the crosspieces, all in a row. It was exactly that simple, but not exactly that easy.

I found in my father's toolbox a gray metal teardrop-shaped thing with a little tag sticking out—a chalk line. My father used to ask me to hold the tag against a fence post, and he would walk out the reel to mark post holes.

"You can fool yerself with too much measuring," he used to say, "but the string never lies."

To create the centerline of my canoe, I stretched the string between both ends of the strongback, lifted the string in the middle, and released it. The string snapped against the two-by-fours and made a puff of blue chalk dust, leaving behind a perfectly straight line. The forms would have to be screwed to the strongback in precise alignment with that chalk line, while also positioned square to the base, and perfectly vertical, or plumb. My father's level was a six-inch-long metal instrument holding two glass tubes filled with fluorescent yellow liquid. A bubble inside the liquid was equal in width to two tiny hash marks on the tube. The center tube read for level on the horizontal plane, like a floor. The other tube measured for plumb on a vertical surface, like a door. This was a silent tool without units: the surface was either level, or not.

Holding the first form vertically against the first station, I gave it a quick tap with my palm. It moved a smidge to the left. I held the level against it and saw that the form was not close to being level. I broke off pieces of shims—flimsy, wedge-shaped scraps of wood—to slot underneath the form. The process seemed no different from leveling a wobbly restaurant table with a sugar packet. The bubble was close, but not sitting between the two lines in the tube. This became a time-consuming process with fractional shifts of the form up and down, left and right. The shim in the corner closest to me got the horizontal reading right, but the vertical surface wasn't plumb. More shims, more minute adjustments, and the form slipped further out of whack. I yanked out all the shims and started from scratch. A shim here, check for level, a shim there, and check for plumb—perfect. Still aligned with the blue chalk line? Yes. Still level? At last, the bubble slipped obediently between its marks.

After attaching the twelve forms with enough screws to withstand a hurricane, I stepped back to assess my work. The forms rose from the strongback like the plates of a *Stegosaurus*, except the silhouettes were uneven. My shaky jigsaw work was exposed, and my first impulse was to carry on and ignore the bumps and gouges. Gilpatrick would have none of that attitude, and he warned against skimping on this phase. The forms are the base upon which I would build the canoe and any lumps on them now would be transferred directly to the hull. To correct this problem, he provided one sentence of guidance: "Inconsistencies in canoe forms can be flattened with the use of a coarse rasp."

I was more familiar with my father's rasp than any other tool in the box. It was stored in a leather sheath that he had stitched together with a veterinarian's suture needle and waxy baseball string. He carried it around rodeo arenas, horse stables, and cattle pastures for so many years that the leather case was worn soft and supple, with scratches and creases marking its journey. Lord knows how many horses had stepped on it.

His rasp was a simple tool with two parts. The menacing business end, a shank of steel with sharp teeth like those on a cheese grater, was made to impose its will. The wooden handle felt warm to the touch, and its anatomical shape was comfortable to grip—however, it was big enough for only one of my hands. If I held it with both hands to smooth out the forms, where would my second hand go? I tried putting one hand on top of the other, thinking they could exert more power together. One swipe on a plywood form and the rasp skittered off. There was plenty of power, but it was like driving a car from the back seat; I could not point the tool from the rear and propel it with both hands. I would have to emulate the mechanics of my father's technique, and I was thirteen the last time I saw him use the rasp at the rodeo.

COUNTRY MUSIC DRIFTED out of Old Yeller, parked with the doors open and a cooler of beer on the tailgate. One by one, rodeo cowboys led their horses up to my father and waited while he trimmed and filed the horses' hooves with his coarse farrier's rasp. He greeted the cowboys with a sheepish smile and a nod, seeming embarrassed by the attention, even though he had been getting it every time he entered a rodeo grandstand his whole life. Farrier work to my father was not something that required payment; it was something cowboys did to help each other out.

To watch him use the rasp was to watch a master in his element. He began by gently coaxing the horse to lift one of its legs and rest its foot on a locust tree stump that he always kept in the back of Old Yeller. This stabilized the hoof so the horse could relax. My father bent over at the waist, gripped the wood handle of the rasp with his right hand and the metal end with his left, and braced his elbows into his thighs. He moved his arms and body in one fluid movement, diagonally across the hoof and back again, his leather chaps squeaking, his breathing in rhythm with each stroke. Hoof shavings sprinkled on the ground like shredded coconut. The younger cowboys gathered around to watch him work, seemingly hypnotized by the repetitive motion of the rasp.

My father explained that each hoof was different, with qualities like wood: varying soft and hard spots that had to be treated carefully to avoid gouging too deep. The daunting physical nature of his work—standing near the muscular haunches of a horse that could kill him with one kick—appeared only to deepen my father's engagement with both the tool and horse. He removed a little material with each stroke and checked the hoof repeatedly to make sure he was not filing off too much.

"I reckon it don't seem like you git much done filing off little bits of hoof, but if you did it all in one fell swoop, the horse would kick yer ass right quick," my father lectured to the gathered cowboys. He took a break from rasping to wipe the sweat from under his cowboy hat with a red handkerchief. "Best to make little adjustments and watch 'em add up."

While my father worked with his rasp, I kept a low profile nearby, roping a plastic dummy steer head mounted to a hay bale. Being inconspicuous allowed me to observe what the other cowboys were doing. Most of them sat in flimsy lawn chairs sipping cheap beer, rubbing baby powder on their saddle straps, and chatting about things like the best flavor of chewing tobacco. They laughed and swore and swatted away flies.

There was a quiet loner cowboy, a bull rider named Jed, who was the closest thing I had to a friend. Jed had ice blue eyes and the sinewy body of an athlete: muscles strung tight like rope, veins coursing down his pale arms. After hanging on to bucking bulls for most of his childhood, Jed's right arm was noticeably bigger than his left. He sported a thick ginger beard and always wore a straw cowboy hat with an eagle feather sticking out. In the triangle of denim below his flashy belt buckle, Jed's endowment was obvious, even from a distance.

The scene unfolded like a film in slow motion, with golden nostalgic light surrounding them. On one side of the horse trailer, my father focused on the task of rasping hooves for a line of waiting cowboys. Jed stood on the other side of the trailer, pressing a wad of beeswax into the worn, slick inner thighs of his turquoise-fringed leather chaps. The wax softened under Jed's fingers into a lubricated smear that he rubbed vigorously until it soaked into the leather. He unbuckled the chaps from around his waist and casually dropped them to the ground. I grew anxious that Jed might peer out from under his cowboy hat and catch me staring at him, but that didn't stop me. His clothes had been so broken down by dirt, sweat, blood, and bull snot that they fit him as a second skin. He peeled off his boots, jeans, and shirt and put on a less dingy outfit—one without underwear or socks, I noticed. For the thrilling few minutes that I watched Jed changing clothes, my whole body tingled with a strange electric pulse that I didn't understand. My heart was racing. Without seeming to notice my presence nearby, or my intense fascination with him, Jed buttoned up his Wranglers, loaded his horse and tack bag into the trailer, and drove away.

Twenty-five years later, Jed was my secret inspiration behind the name of our new red wine, First Crush.

There was a time after my father's funeral when I paid Jed a visit. I found him working in his barn at the end of a long gravel road. Jed never left South Dakota, and the irony was not lost on me that he had more in common with my father than I did. Jed and I chatted about horses and rodeos, college and car accidents, and dogs we had both owned. He finished working and invited me inside his cramped mobile home for a drink.

The sex was not something we talked about, or something I expected, but it unfolded naturally after a few beers.

"You know I ain't queer, right?" Jed said afterward. I got dressed and drove back to New York.

I SHIFTED MY top hand forward, grasping the end of the steel shaft while leaving the other hand on the wooden handle. I pulled the rasp across my body in a long, diagonal stroke. Instantly I gained control over the tool. Again and again, I filed off a little material and checked the surface of the plywood canoe forms. By this slow, careful process, I adjusted the edges right up to the original pencil lines I had traced from Gilpatrick's patterns. Fine wood shavings the texture of grated coconut accumulated on the floor as I moved down the length of the strongback. Sometimes the rasp dug in too deep and stopped. Other times it slipped over the plywood and my wrists cuffed the edge, leaving traces of blood. As the rasp moved across the wood, my whole body moved, too. My technique wasn't smooth or confident like my father's, but it was competent enough to get the job done.

I pressed my face into the leather sheath of the rasp and breathed in slowly, hoping for the faintest smell of the rodeo. There was no real scent, though, only the memory of it—and the imagined romance of my former cowboy life, of which nothing was left but what I held in my hands.

CHAPTER 18

BRANDING DAY

The postmark read Washington, D.C., and the return address was stamped in black boldface type: U.S. National Archives and Records Administration. Contained in the envelope, I hoped, would be answers from forty-five years ago.

Until the letter arrived, I had completely forgotten that I mailed a next-of-kin information request for my father's military service records the day after his funeral. Maybe this would be a breakthrough, a chance to gain clarity about his time in Vietnam or any insights as to why he received the Bronze Star. I opened the envelope and carefully removed the letter.

Dear Dr. Preszler . . . in response to your request . . . no further information is available about your father's military service . . . such records either do not exist at the National Archives or remain classified. . . .

I stared at the letter, disappointed, but not surprised. It was hard for me to imagine ever having a complete picture of my father.

Everything I knew or did not know about him just became more complicated, but strangely simple, too. There was no explanation. I would never get an answer. It was time to move on.

I gathered an armful of offcuts and started a fire on the beach. The blaze was high, wide, and orange when I tossed in the letter and watched it turn to ash. Caper and I stayed on the beach longer than usual, sitting by the fire well after night set in. The bay was flat calm with no sounds of waves or water. The landscape was visible only as darkness and lighter darkness. Near the fire pit, I caught occasional glimpses of the *Spartina* cordgrass, scrub oaks, and junipers that clung to the dunes—now all the same piece, blacked out, unimportant, and shapeless.

Looking back toward the house, I saw the windows glowing amber from the filament bulbs I had strung across the ceiling. Nearly every square inch of floor space inside was covered with lumber, tools, boxes, and the canoe strongback. When I was standing right next to the strongback, fussing with the level, chalk string, shims, and rasp, I couldn't perceive the overall shape or magnitude of it—a case of missing the forest for the trees. Seeing it for the first time from some distance away, though, it bore the unmistakable silhouette of a canoe in skeletal form. It looked like something that might someday be a real canoe! This revelation filled me with great, unbridled hope.

A few seconds later, all hope dissolved when a different kind of revelation sank in. The canoe skeleton was not just big and long; it was outrageously huge. It filled my entire house from one side to the other—a sea monster big enough to swallow me whole. What on earth was I thinking by building this? What if this giant wooden contraption was not *just* a canoe? I had never thought of it this way before, but what if the canoe were the physical manifestation of my grief—and not just grief, but the unresolved angst that parts of my father would always remain unknowable.

I dragged myself into the house and called Mom.

"Hello!" she said.

Silence.

"Hey, Mom."

I stood still, unable to walk around my own living space.

"How are you?"

"Fine, Mom. How are you?"

"I'm okay. Working on a quilt."

A pause.

"I heard back from the National Archives about Dad's Vietnam service."

"And?"

"They got nothing. We got nothing. There's just . . . nothing."

In the silence that followed, I couldn't hear Mom's pain over the phone.

THERE WAS ONLY one occasion in my life when I heard my father speak the word *Vietnam* out loud. I was ten years old and it happened on branding day, which was practically a Preszler family holiday. Whereas on the other three hundred sixty-four days a year I was alone with my family at the ranch, on branding day a dozen other trucks and horse trailers pulled into the yard at sunrise. Without much conversation among them, the neighboring ranch families helped us round up the cattle herd, a thousand animals registering loud, guttural dissents. Dust encircled the procession as we moved toward a chute opening in the corrals. Cows mooed. Calves bleated. Cowboys barked and whistled in the morning light. My father clicked his tongue and made kissing noises to get the cows' attention and yelled when he needed them to pick up the pace.

"GIT!" he shouted. "HAW!"

Inside the corral, my father's hunting buddy Dwayne separated the young calves from their mothers while my father dug a hole in the ground to heat a branding iron with a propane torch. Our family's brand was called T-lazy-J and it marked our livestock in case any escaped or were stolen by bandits. Beyond its practical purpose of identifying our property, the brand was also a link to our past on this land: it was registered with the South Dakota cattle commission at the turn of the century.

The school-age sons and daughters from neighboring ranches lassoed calves by their hind feet and dragged them on their sides across the dirt to the other end of the corral, where Jed and a team of older high school jocks wrestled them to the ground. We scrambled in like pit crews at a racetrack. My grandfather vaccinated the calf, Mom pierced its ear with a yellow plastic identification tag, and my father slammed the red-hot branding iron against its rump. The hide sizzled, smoked, and flamed. The air, already swirling with dust and noise, filled with the putrid scent of burning hair and flesh. The calf wailed. After a few seconds, my father lifted the brand, and the calf would be scarred for the rest of its short life.

My role in this mayhem was to follow my father around carrying a pail of ice water, known as the nut bucket. Castration was quick. My father tugged on the calf's scrotum and sliced away its testicles with his hunting knife. He dug his fingers inside the body cavity and pulled out bands of tissue connected to the testicles and sliced them away, too. Without turning his head to look, he flicked the pair over his shoulder, and I caught them in the nut bucket. Sometimes I missed, and the slimy, bloody balls slapped me in the face, still warm, and stuck there like masking tape. When my father finished, the calf clambered to its feet and ran away, reuniting with its mother in the freedom of the prairie. This process repeated hundreds of times over the course of the day, the dust and cacophony rising over the corrals.

When it was over, Mom and the other ranch wives—who called themselves the Cow Belles—unveiled giant tubs of potato salad, deviled eggs, and the greatest western delicacy of all: Rocky Mountain Oysters, or calf testicles fried in salted butter. Eating bull balls was a cowboy tradition to celebrate the end of branding. Guys took turns tossing fried balls in the air and catching them in their mouths. I tried one and spit it out. The rubbery texture, and the musky, animal funk triggered my gag reflex.

After all the calves had been branded and the neighbors went home, my father took me into the pasture, driving Old Yeller at a snail's pace across the bumpy prairie. The truck jolted back and forth as its suspension and chassis creaked in protest. The sun was bright, the sky crystal blue. The smell of sweet clover filled the cab. My father, sweating through the sleeves of his thin, snap-up Western shirt, hung his left arm out the window. His thick hands were tanned deep brown.

We came across a dozen cows chewing their cud alongside a bull named Herman. I had secretly named the biggest and oldest bull in our herd, even though my father warned against naming any animal that could end up as supper. Herman weighed almost three thousand pounds and moved with a long gait and unspeakable power. He used to lumber up to the fence by the schoolhouse and wait for me to feed him carrots. He would slurp one from my hand with his pink sandpaper tongue and lower his head so that I could scratch the curly white hair between his eyes.

My father stopped the truck about ten feet away and jotted notes about each animal on a small legal pad. He called this activity checking cows, and it was a routine part of our summer ranch chores. On this occasion, however, something happened that I had not seen before. Herman reared up on his hind legs and mounted a cow that strained under his weight. I watched in shock and confusion as a three-foot-long, slimy pink dart slid out of Herman's underbelly and disappeared into the back side of the cow. A couple seconds later, he was down on all fours again, chewing grass and swatting flies like nothing had happened.

"What's he doing?" I asked with wide-eyed disbelief.

"Breeding," my father said, without further elaboration.

In the distance across the prairie, an ornery two-year-old bull sprinted toward us. It was the same bull that had rammed the horse trailer and smashed my fingers the previous summer.

"Here comes trouble," my father said.

He honked the horn and snapped his bullwhip out the window to warn the herd, but they paid no attention. The young bull continued rumbling in our direction, closing the gap with every thunderous stride. At the last second, the bull swerved past Old Yeller and rammed into Herman while he was in the act of breeding another cow.

Upon impact, a thick skull driven by several tons of momentum met a hollow rib cage, and Herman let out a monstrous bellow. He toppled over and landed with a thud in a cloud of dust, kicking and flailing and twisting his neck, trying to rise again.

"Herman!" I howled.

"Goddamnit," my father said—not the halfway curse *galldangit*, but the real thing.

As Herman pulled himself up onto his front legs, his back legs buckled in excruciating slow motion beneath him. He tried to take one wobbling step forward by scraping his front hooves on the ground, but his body tilted hideously backward, as if he were the *Titanic* sinking into the sea. In all the commotion, the rest of the herd ran off with the younger bull and left us alone on the prairie with Herman. He hovered there and let out a deep moan that sounded like he was beseeching me to help.

My father loaded bullets into his .30–06 Winchester and scrambled out of the truck while unleashing a flurry of curse words. He stomped toward Herman and faced his massive head at eye level.

"What are you doing?" I asked from the safe confines of the truck.

"I gotta shoot him," he said, pointing the gun at Herman.

"Why?" I screamed back at him.

"Cuz that sonofabitch broke his back and if bulls don't breed, they're worthless."

My father fired the gun and I inhaled sharply. The bullet hit Herman in the patch of curly white hair between his eyes. He sat unmoving, looking at us with a stunned expression. My father fired another bullet into his head and one into his chest in quick succession. Herman jerked backward but did not fall. His eyes were wild upon us and his body was riddled with bloodless holes. I had shot deer with my father before and had a pretty good idea of what it took to kill a four-legged animal. Nothing could have prepared me to witness the takedown of a three-thousand-pound bull. I should have known there was no such thing as a clean kill.

"Shoot him again, he's not dead!" I begged, shocked by the guttural sound of my own voice. I buried my face in my hands, unable to watch.

"DON'T LOOK AWAY!" he hollered.

I uncovered my eyes.

He fired two more bullets into Herman's head. Blood gushed from his soft pink nostrils in a sudden, great torrent, splashing onto the sunbaked prairie. He coughed and coughed, more blood pouring out with each heave. He struggled to stay up before his front knees buckled grotesquely underneath him and he collapsed onto his chest.

I staggered out of the truck and ran to him. I placed my hands on his rib cage. It moved up and down, slowly. He breathed enormous breaths and then finally dropped his head to the ground. I felt with both of my hands the moment his body went still.

My father removed his knife from the leather holster on his belt. With one violent upward sweep of his arm, he cut off Herman's ear to save the yellow plastic identification tag. Blood splattered his snap-up Western shirt.

"Just like Vietnam," he said.

IN THE YEARS that followed Herman's bloodshed, that day's carnage did not haunt me as much as my father's words: *If bulls don't breed, they're worthless.* The more I considered them, especially as I went through puberty, the more I worried that I might be maturing into the wrong kind of man. By the time I was sixteen, my body was changing, swelling, and stretching in new ways that I quite enjoyed. The only problem was that the more my body changed, and the more I became a man, the less I had in common with men. Girls were practically invisible to me. Sometimes I would be in the locker room after high school gym class and I would become paralyzed, scarcely able to walk or talk because I was so preoccupied with what I saw in the showers. Worse yet, I feared the consequences of getting caught looking or saying the wrong thing at the wrong time to the wrong guy. I had learned at summer Bible camp that even if I abstained from having sex, I would still commit a heinous sin every time I merely *thought about* it. Even my innermost thoughts that I could not control were an abomination.

Not being into guys was the thing I prayed for most often, whenever I remembered to pray, but it was a futile wish. My body did not seem to care what my mind thought about my strange attractions to men, or how society and the church punished gay people. My body belonged to itself and I was destined to become, in my father's language, one of those worthless bulls that didn't breed.

AFTER DOUSING THE beach fire with seawater, I crawled into my sleeping bag next to the strongback, and I dreamed of blood: the blood that was spilled when my father had shot Herman; the blood of the nameless enemy soldiers that I could only assume he had killed in Vietnam; the blood that dripped on the wood when I had first used his handsaw; and the blood left behind on my face and hands after helping him castrate calves. The memory of most of those things was more frightening than the actual experience of them had been.

The next morning, I took my father's hunting knife out of his toolbox and used it to cut a slab of bacon that I fried—extra bacon for Caper, too. As the blade sliced through the muscle and fat of the pork belly, I saw that I didn't have to be frightened. My father's knife was not dripping with blood. In fact, it was an elegant piece of handcraft. He had forged it himself by heating up an old wood file and honing it down to a six-inch blade, so that the cutting face was smooth, but the thick top edge was grooved from its previous incarnation as a file. He had outfitted the blade with a two-toned walnut handle with finger notches that mirrored the contours of my right hand. Being so well balanced, and so suited to my hand, it felt lighter than it looked.

All morning, as I studied the next steps of canoe construction, I thought of the trusted knife that had clung to my father's right hip his whole life. I did not think so much about the knife itself or how he used it, but more about the family heritage that it represented. I spent the next few hours reading nautical history books, fantasizing in colorful detail about what it would be like to paddle my own canoe. I felt a tug inside me that was related to connecting my past with my present, to squaring up where I had come from to where I lived now. Since starting the canoe, I had become more aware of that bridge between my two selves, and that awareness blunted my old ways of thinking when I tried habitually to sever those connections.

There was an important commonality between the horseback-riding cowboys of the Great Plains and the canoe-paddling Iroquois tribes of New York. Both canoes and horses were the means to explore new territory, trade goods, and gather food—a different medium with the same net result of getting from point A to point B in a landscape without roadways. While cowboys branded their cattle, the Iroquois imprinted family symbols on their canoes, either with a mixture of pine tar and pigmented herbs, or by burning them into the wood. I had first read about this practice in one of my history books, but it was not until now, a sunny day in mid-April, that the similarity between the two traditions crystallized in my mind. What the Iroquois knew when they had emblazoned symbols on their canoes was the same thing that my great-grandfather, grandfather, and father knew when they used the Preszler family's branding iron. They were not just marking property; they were passing on an heirloom that linked them to their past, to their family bloodline, and to a shared purpose of building and growing things. It did not matter so much that everything about my sexuality and my canoe would have been foreign to my father, or that his Vietnam service would remain an unsolved mystery. What did matter between us was utterly timeless. We were of the same blood, and that was one bond neither of us could sever.

I sent Mom a text message.

"Do you still have our cattle branding iron?" I asked, a little nervous that I had waited too long, and it would already have been lost to the garbage collectors or given up at a rummage sale.

"Oh, sure! It's in the garage. Dad made sure we kept it," she replied.

"Could you mail it to me? I want to brand my canoe."

CHAPTER 19

BREAKING POINT

The first time I used a table saw, the blade hit a knot in the juniper and the entire board shattered. Shards exploded back toward me like bullets. One piece caught me in the thigh, narrowly missing my crotch. A bigger piece rocketed past my head and punched through the sliding glass patio door, showering thousands of glass bits onto the tile floor. For a few dazed moments, I stared in shock at the broken glass while cool spring air flowed in and Caper cowered in the kitchen. I swept up the glass and felt like an idiot for ignoring the safety manual. When I brought the table saw home from the hardware store, I simply unpacked it, plugged it in, and started sawing, counting on its operation to be self-explanatory. I had watched my father use a table saw without incident hundreds of times, but that was exactly the point. It had taken him years of practice to make dangerous tasks look effortless.

What I had been trying to do with the table saw seemed so basic in theory. I needed to rip-saw all my lumber into quarter-inch-thick strips that would be molded around the strongback forms. Each new layer of strips would gradually build up the boat and close the hull.

Molding thin wood strips around forms was not the only way to build a canoe, just the current way. Woodworking tools used for carving and shaping dugout canoes—hollowed-out tree trunks—had been found in coastal Maine and carbon-dated to four thousand years ago. Over time, Native tribes abandoned dugout canoes and began crafting them out of thin strips of birch bark that were lashed together with spruce roots and sealed with boiled pine tar and bear tallow. In modern times, dugout and birch bark canoes were replaced by those made from wood strips and sealed with fiberglass, like the one I was building.

After buying a Plexiglas face and body shield to protect myself from flying shrapnel, I carried on sawing with some trepidation. I adjusted the metal guide that ran parallel to the blade so it would rip a quarter-inch-thick strip off the board in the long direction of the grain. The table saw roared to life like a jet engine and I pushed a board through the whirring circular blade with a plastic stick that kept my appendages a healthy distance away. Rust-orange sawdust spewed out the back exhaust. Watching the jagged-toothed blade spin on its axle, I feared that it might shoot out of its slot in the metal table and slice through my guts, all in the blink of an eye. Every time the roar of the saw changed pitch, I flinched, aware of the damage it might inflict. It took a leap of faith to continue sawing, but I did.

Two weeks trudged past with the spinning blade wreaking havoc on my self-esteem. Aside from hiring a professional to re-glaze the broken patio door, I didn't accomplish much. I would come home from work every day, mangle a few strips, and go to sleep. The slow-motion lack of productivity was anxiety inducing. Every third or fourth strip exploded when the blade hit a knot. Most of the lumber was so riddled with knots that I would get only two usable strips from a six-inch-wide board. The rest shattered on impact or left me with half strips, three or four feet long, which were not of much use. I regretted that I had ignored Scott's recommendation of using soft western red cedar instead of its brittle cousin, juniper. I was in so deep now that there was no turning back—not when I had bought so much of it.

Sometimes the wood cut perfectly along a clean line that was straight as an arrow with a gorgeous deep red grain and a spicy cedar aroma. Other times, the strip peeled off the mother board in a hideous warped curve that crumbled upon itself. There was no way of knowing which it would be. Every board I cut was a game of chance, a coin toss.

I tried cutting strips from shorter boards and longer boards. I cut strips during warm, humid days and cool, dry nights. I fed the boards through the table saw slower, and faster. I lowered the blade height to just above the level of the wood. I raised it. I lowered it again. I wore different clothes. I soaked the wood in the bathtub. I changed saw blades. I tried everything to fix the problem, then came to believe that the problem could not be fixed because the problem was me.

Every night after another frustrating battle with the table saw, I scribbled notes in terse shorthand on the chalkboard wall. My notes were either a form of journaling or an avoidance mechanism, or both, and helped me get a grip on what my strange new life was all about. I tallied the number of usable strips I had cut. I crossed off each passing day on the calendar with an *X*. I copied useful tidbits and inspiring quotes I had gleaned from books. I traced hull blueprints and ranked the relative hardness of tree species commonly used in boatbuilding: teak, mahogany, and purpleheart. Eventually, my chalk wall notes evolved so that they were no longer just about the canoe and wood. I jotted down quips about the weather (*freak storm squall: double rainbow!*); I described the various states of Peconic Bay (*slate gray waves moving parallel to shore without breaking*); I observed the seagulls (*What's the one with the black head and yellow eyes?*); and documented Caper's eating habits and heart medication schedule. While I struggled to make sense of my failure to master the seemingly simple act of cutting wood strips, the walls filled up with the hieroglyphics of my daily life.

One weekend morning, I dragged myself to lunch under a rainy spring sky the color of wet concrete to meet an old college friend, Brad. He spoke at my wedding in Spain two years before, and I had done so at his wedding on Cape Cod years before that. We had played tennis on cracked asphalt courts in Iowa in the 1990s, danced at gay clubs in Greenwich Village in the 2000s, and hiked to the rim of a volcano in Patagonia in the 2010s. He was passing through New York that weekend and wanted to meet. We sat down for lunch, our first real time to catch up since the events of my life had taken a turn for the worse. I knew I should talk to him but couldn't remember how. I tried some words, but they felt wrong. I was exposed. There was no way to hide my depression. Back at the house, I showed Brad my canoe and he surprised me by saying that his father, Dean, had also built a wood strip canoe. He encouraged me to visit Dean, who lived in Ames, Iowa. By coincidence, I was scheduled to attend an alumni board meeting at Iowa State the following week, so I did.

Dean took me into his basement and showed me the wooden canoe that he had built from scratch. It was a magnificent work of art. I had never seen one in person before and couldn't take my hands off it. I caressed the sensual, curvaceous cedar hull and admired the varnish, which was so glossy that it was almost blinding to look at. Dean reached up and pulled out from the rafters a twenty-foot-long, straight-grained piece of western red cedar that had been tucked behind a tarp. It was the perfect canoe strip: no chips, no cracks, and a consistent thickness. It was so pure that it might as well have been carved from butter. Dean had such a gift with woodworking that he kept extra strips hanging in his rafters. I couldn't believe this: *extra strips!* I had been sawing wood for almost a month and had not achieved even one piece as remotely attractive as the gorgeous gems that Dean had stockpiled. He slid the strip back into its ceiling rack and, inch by inch, the cedar disappeared. I was left staring in awe at his expertly built masterpiece.

Dean offered me a clamping jig that he had made to streamline the process of joining multiple short strips together into one long strip. "Would you like to have it?" he asked as if he were passing off a Christmas sweater to a poor relative. As I weighed the clamp in my hand and studied Dean's drawings—which would allow me to replicate the design and make more clamping jigs at home—a wave of humiliation slammed into me. I was accepting charity for a task that I should have been able to figure out on my own. Nonetheless, I took it gladly, of course. I needed all the help I could get. I told him that I was relieved to have his help, that I felt somewhat desperate, that the work on my canoe had been frustrating, and that things had been hard since my father died. I would have liked to stay longer, but I had to catch a flight in Des Moines. He nodded in understanding. I took the clamping jig and left.

On the plane, I couldn't stop thinking about those perfectly milled cedar strips hanging from Dean's ceiling. I wanted to travel back in time and push past Dean and tear them down from the rafters and steal them, to run screaming down the street holding them over my head like a trophy. While I felt admiration for this nice, honorable father and craftsman who loved his son, one of my best friends, to my great shame, I hated my reaction to his canoe. It embarrassed me to acknowledge that I was jealous—not only of Dean's expert boatbuilding skills, but also of Brad's supportive and healthy relationship with his father. It would have been uncouth to let those feelings surface in any meaningful way, though, so I ordered a cocktail and slept the rest of the flight.

A couple days later, I crashed my car on the Long Island Expressway when a deer bounded out of the woods. In the millisecond I had to react, I kept my hands on the wheel and barreled straight ahead, bracing for the impact. We collided with a sickening thump, and the crunching of antlers and bone with glass and steel. I was not injured; the deer was dead. At the BMW dealership where I had bought the car days before my first cross-country road trip, the service technician joked, lightheartedly: "I remember when you picked this up before Thanksgiving. How did you manage to rack up so many miles in such a short time?" I thought about his question and knew it was rude to project feelings about my father's funeral onto this innocent mechanic, so I checked my worst impulses.

"It's such a funny thing," I said with a half-hearted laugh. "Would you believe I drove across the country, twice?"

I finally got home with a loaner car and was back in the office the next day when a text message arrived from Mom: "Did you get the cattle branding iron yet? FedEx tracking says it was delivered last week, but I haven't heard from you."

I hadn't received it, even though FedEx confirmed that it had been delivered to the winery along with dozens of boxes filled with wine labels, glass bottles, and corks. The operations director, Peter, searched the winery from top to bottom but the package from Mom was nowhere to be found. I drove to the FedEx depot and waited in line for an hour to speak to a customer service attendant who was politely doing her job when she asked me to fill out a loss claim form.

"Sir, I just need you to describe the contents of the box and estimate its value," she said.

I took a deep breath to collect myself. Of all the packages I had sent or received in my lifetime—most of them containing cheap crap from Amazon that was made in China—this one was truly irreplaceable: a symbol of traditions from a life long gone. How could I possibly estimate the value of a priceless family heirloom?

"Sure, no problem, I'd be happy to fill out the form. Let's see, maybe I'll put down a hundred bucks or something," I muttered while scribbling illegibly on the form, my hands shaking. "I understand things get lost sometimes, it's one of those unavoidable facts of life, right? Can't do much about it. I know it's not literally *your* fault."

On the drive home, I called Mom with the bad news. We tried to guess how it might have gone missing: perhaps it was mistakenly thrown away with the cardboard, or was stolen for scrap metal.

"Who else in New York would even know what a cattle branding iron is, much less see value in it or want to take it?" Mom asked.

"Nobody."

I slumped into my chair at home surrounded by mountains of shattered juniper strips that resembled landscaping mulch. All I wanted was for this canoe build to go smoothly and to incorporate a little token brand from the ranch, but even that uncomplicated idea blew up in my face. Despite my constant attempts at controlling this process, making sense of my father's death, and creating order out of the chaos of my life, with every step forward I took, I introduced more uncertainty. I blamed my father for getting me into this mess in the first place, and I blamed myself for believing that I could accomplish such a crazy feat as building a boat. The world with the canoe in it was supposed to be insulated from harm, and in that world, I was supposed to find my destiny, to learn of my limitless potential as a boatbuilder. Instead, I learned just how limited I really was.

Every day for the next two weeks, I walked and talked and slept and dreamed of canoes, of wood and saws, of paddling through waves under flocks of migrating ducks. Every night after work I came home to the canoe strongback with relief, because when I was there with it, I knew better who I was, and some of my frustration melted away, even if I wanted to weep on my knees every time another strip shattered.

On my birthday, I gave myself a gift: I would attach the first strip to the strongback forms. After so many weeks of toiling with the table saw, I had to break the monotony. The strips I had cut were too short, but I could use the clamping jig that Dean gave me to join several short pieces together with tapered, overlapping ends, called scarf joints. If, by some miracle, I successfully created a twenty-foot strip, the run of its grain would follow the curvature of the canoe. The act of making these curves would be the first time I could exercise any degree of artistic flair, coaxing the wood into a shape it would not normally follow. There was an inherent tension in bending wood, and until it relaxed in position and conformed to the curves, it would have a natural tendency to spring back to its original shape. To mold the wood in the shape of the canoe would be to tame the natural stresses within it and minimize its stiff internal tensions.

I marked off narrow angles and butchered the ends of the short strips with my father's handsaw. The scarf joints were supposed to look like thin triangles, which would fit together like puzzle pieces inside the clamping jig. My scarf joints were clearly jagged and mismatched from the start, but I didn't want to waste time worrying about them. I was in a hurry and desperate to force real, material change to happen. I held a short strip segment to the form amidships, lined up a tiny nail and drove it through with my father's hammer. The flat face of the strip fit tight against the form. It felt exhilarating. I let out a sigh of relief. At last, something went according to plan. I clamped the short strips end to end with the jig and dabbed wood glue into the mismatched hodgepodge of scarf joints. Then I wrapped blue masking tape around each joint to hold it while the glue dried. Once finished, I repeated those steps on the opposite side of the canoe, so I had one strip on each side, port and starboard. Where the two strips met in points at the bow and at the stern, I glued and clamped them together. This felt monumental. All other strips of the canoe would be built on top of these first, baseline strips, called the sheerline. It was like laying the first row of a brick wall. There would be no other curve on the canoe, and no pieces of wood more important than the sheerline strips I had just nailed and glued in place.

I inspected my work with one eye closed, squinting down the length of the strongback. The curve was not entirely smooth, but I was already months behind on this project and it would have to do for now. Less than an hour later, I was folding laundry in another room when Caper started barking in the living room. I walked over to investigate.

Pop!

Caper kept barking.

Pop! The sound came from the canoe. I looked closely. The glued scarf joints between the short strips were failing, one by one. It was not because of Dean's jig; that was perfect. No, the joint failure was my fault. The strips were of such wildly inconsistent thicknesses that the scarf joints didn't align. I had also neglected to fully tighten the clamps that held the two surfaces together while the glue dried. While I had intended for the strips to lie like onionskin against the forms, they bubbled out at each of the slapdash scarf joints.

Pop!

The strips snapped back to their natural straight shape, and any semblance of a graceful sheerline curve disappeared. Shorter strip segments simply broke apart and dangled from the strongback by feeble pieces of blue masking tape. Others crashed to the floor.

Pop!

I wanted desperately to snap my fingers and have the canoe finish itself as if by magic. I wanted my canoe to look as beautiful as Dean's canoe had in his basement, but I knew it would not. It will be fine, I told myself, I'll adjust these strips here and tape them together over there—

Pop!

The torque from bending the wood around the forms caused the tiny nails to dislodge. Caper was now barking nonstop, losing his mind over this twenty-foot noisemaking sea monster that had taken over our house. I cursed. Maybe if I just added another dollop of glue that would fix the—

Pop!

That was it. I hated this damn canoe. I was outraged by its existence. I felt a blind, cold, shaking rage rise within me. I grabbed my father's hammer and started swinging—big, indiscriminate swings. First, I struck the sheer strip at the bow. It shattered and fell to the floor. Caper ducked for cover in the kitchen. I slammed the hammer down on the biggest mushroom-shaped form amidships, knocking it out of plumb and off level. I felt the frustration of being in over my head, hoping I could find some angle that made sense of this canoe, but I couldn't. I struck the strongback with the hammer as hard as I could, many times in rapid succession, grunting louder with each strike. The two Oregon Douglas fir beams that formed the frame of the strongback made awful, high-pitched splintering sounds. One of the beams cracked. I punished it again and again. All the rage that had built up inside me over the previous six months, all the resentment I did not know was there, reared up in a towering instant of fury. The crack in the beam that started ten rings deep opened outward and split off the end of the strongback, taking three of the forms with it. The strongback collapsed in a twisted and hideous heap on the floor, like Herman the bull's broken back. The house fell silent.

I opened the patio door, ran out to the beach, and hurled the hammer into the air. I let out a mad howl.

"FUCK YOU, DAD!"

His hammer tumbled end over end and landed with a plunk in the sea.

CHAPTER 20

VALLEY OF THE GIANTS

If I drew a map to illustrate my schedule over the following month, it would be an explosion of lines emanating away from New York City. I went to Rochester, Syracuse, Ithaca, and back. To Washington, D.C., and back. To Chicago and back. To Salt Lake City and back. To Los Angeles and Seattle and Portland and back. I travelled for meetings as chairman of the New York Wine & Grape Foundation and chairman-elect of WineAmerica, the national association of American wineries. I had only been elected to these positions a few months before my father died, but my interest was already waning. I had overcommitted myself to professional causes that no longer inspired me, and I was unsure if they had ever inspired me to begin with. The ambitions and life plans that once fueled my success had faded, replaced by a weird lack of motivation. I couldn't tell if that meant my frustration with the canoe was spilling over into my professional life, or if I would rather be at home working on the canoe—or both.

The last of these business trips was a corporate leadership retreat for chief executives at a winery spa in Oregon's Willamette Valley. I was walking through the spa's lobby when I saw a Bureau of Land Management Oregon-Washington brochure in a plastic rack near the concierge desk. I stopped briefly to read about a pocket of old-growth forest nearby called the Valley of the Giants. Among hemlocks, cedars, and other monster conifers, one of the world's largest Douglas fir trees lived in the valley and had somehow escaped the chainsaws of loggers for the last five hundred years. I tucked the brochure into my briefcase and carried on to the conference room, where I sat through an unbearable three-hour roundtable discussion about alcohol excise taxes and public-private growth synergies. I couldn't focus. I loosened my necktie and urged myself to concentrate on why I was there, to remember that I had a business to run and a reputation to uphold. I was often torn between different places and identities that somehow prevented me from feeling too comfortable in my surroundings: too corporate for woodworking, but too crunchy granola for board meetings. All I could think about now, though, was the forest.

Ever since I was old enough to help my father set juniper posts in the ground for cattle fencing, I loved and respected wood as a living organism. I didn't think trees had a consciousness, per se, but rather an innate responsiveness to natural forces like wind, rain, sunlight, geological time, and the gravitational pull of the moon. Despite trying to build a wooden canoe and holding trees in such reverence, I had never actually visited an old-growth forest and stood among the giants. I was here in Oregon, the center of American forestry, and this was my best chance to experience it firsthand.

I sneaked the Bureau of Land Management brochure out of my briefcase and studied it underneath the conference room table. Spanning fifty acres deep inside Oregon's Pacific Coast Range, the Valley of the Giants was a pitiful remnant of the old-growth forests that once cloaked the entire continent. The place held some mystique among Oregon locals as few people had ever seen it. The approach to the trailhead was apparently one of the most difficult that a person could ever endure: thirty miles of single-lane, steep, twisting, unmaintained, and unnamed logging roads through private timberland. For most of the year, the route was impassible due to snow cover, washouts, and falling boulders. None of that deterred me. I stood up and walked out of the meeting with my phone pressed to my ear, faking that I had just answered some urgent call from the winery. But there was no voice on the other end. I would not be returning to the dreadful growth synergies roundtable discussion, nor would I be showing my face at the last two days of the executive leadership retreat. The forest was calling, and I had to go.

I scrambled back to my hotel room and shoved everything I had unpacked back into my suitcase without the usual orderly care for what went where. I was in such a rush that I didn't change clothes, not that I had anything appropriate to wear anyway: all I had packed were business suits. I drove southwest on Route 99W and zipped through the mountains on several paved highways until I reached a gravel road that twisted and turned through valleys, over small creeks, along ridges with sweeping views of the Pacific Coast Range. It was an overcast day in late spring, and the road was lined with wild irises, orange Indian paintbrush, and lavender lupines that huddled in microclimates created by slight changes in elevation and sun exposure. At some point, my botanical joyride was interrupted by a nagging feeling that I had been driving for too long. Without a cell signal for directional guidance, I pulled over to check the folded road map I had purchased at a gas station on Route 99W. Sure enough, I had been so distracted by all the plant diversity that I had missed a turn and overshot my destination by several miles to the south. I used the paper map to navigate my way back north, taking some anonymous private logging roads deep into the mountains. Eventually, I came to a gate marking the entrance to the ghost town of Valsetz, among the wettest places in the world, receiving up to three hundred inches of rain per year. For now, there was only a light sprinkle on my windshield.

For two hours, I crept along a twisting, potholed logging road without seeing another car or person. The road had practically split open in some places, with streams of water flowing in wide gullies across it. I came to a Y junction near a sea of aspen trees. The air was dead calm, but the aspen leaves fluttered as if there were wind—their long, flattened leafstalks twisting at the slightest change. They sounded remarkably like the cottonwood trees in South Dakota: the fluttering of distant applause. The road dropped down a steep, eroded track into a mossy rainforest of Douglas fir and western hemlocks shading patches of salmonberry, sorrel, and trillium. I noticed some very large trees—larger than anything I had seen in my life. Down the slope, the road ceased being a road and became more of a gravelly tire track through elderberry thickets. The potholes grew deeper and muddier, and my rental car proved inadequate for the terrain. The tires spun out. Ahead of me about a hundred yards, a fallen tree blocked the track. The sun faded to a copper penny in the western sky.

I was still wearing a business suit, and my feet sank in mud up to my ankles. I didn't know what I should do. It was twilight, plunging fast toward full darkness. The only thing more upsetting than the thought of walking an unknown logging road in the dark alone was not seeing an ancient tree. I had plenty of bottled water and gas station snacks, so I hunkered down for the night. I would hike the rest of the way in the morning. I reclined the driver's seat and lay back in the dark. The car's roof came alive with the assault of rainwater.

MY FAVORITE SONG growing up was an old country tune about rainy nights. During bad weather, I used to tack a blanket over the corner of my bedroom to make a canopy, under which I would play music and stage races with my Matchbox cars. Lucy and my Muppets were the only race spectators allowed in the room. We would choose songs from a collection of 45-rpm records that Mom gave us. The room felt lifeless without the turntable playing. One time, I pointed to the stack of records and asked Lucy which one she liked best. She pulled one out and stared at the label.

"Me no yike it," she said.

"Which record *do* you like?" I said, shuffling the black vinyl in my hands and fanning the discs out like a deck of cards.

"Wh-which one Tent yike?" Lucy said.

One record cover had a close-up photo of a bearded man with a shaggy 1980s hairdo and wearing a satin jacket over a button-down shirt with its collar popped. I pointed to it. Lucy squealed in delight.

"F-f-f-avorite!"

I clipped the vinyl into the turntable and gently sat the playing needle down on its surface. The record scratched to life and the song filled my bedroom: "I Love a Rainy Night," by Eddie Rabbitt. I squatted on the floor by Lucy's wheelchair, racing cars around my imaginary track. After a while, I stopped and lay next to her feet, wool tube socks covering her orthopedic shin splints. I listened to the song. We hummed and half lip-synched the familiar lyrics together.

I love a rainy night.
I love to hear the thunder, watch the lightning
When it lights up the sky,
You know it makes me feel good.

It was a celebratory song with a beat that made me snap my fingers. While the lyrics may have been a metaphor about finding goodness in dark times, as a kid, I took them literally. To me, it spoke about being trapped but safe inside during a thunderstorm—and we grew up with plenty of those. Every summer, heavy air masses from the Rocky Mountains clashed with the Gulf Stream over the flat prairie to create powerful storms that shaped our lives. Hard sheets of rain could turn to hail in a matter of seconds and destroy the year's wheat crop. My father didn't worry much about the storms until the air became still and the cattle ran in from the pastures to huddle against the barn. Then we knew things were about to get interesting. If a tornado seemed imminent, we would ride out the storm in the basement, a damp space lined with shelves of Ball jars filled with pickled vegetables and chokecherry jam, covered in gray dust and cobwebs. We would play the card game Uno by candlelight until the rain stopped, then emerge upstairs to see what kind of havoc the Lord Jesus had wreaked upon us. A grain auger would be flipped on its side, or a dead cow would be floating in the stock tank, struck by lightning while she was having a drink.

On my thirteenth birthday, my father phoned in a request to the local AM radio station. I was standing on a stepstool licking birthday cake batter off Mom's mixing spoon when the announcer said, "Here's a little song going out to Trent Preszler on his birthday, 'I Love a Rainy Night,' by New York City's very own Eddie Rabbitt." I whooped and hollered and ran around the house, pushing Lucy's wheelchair in circles while she giggled. Anything in the world felt possible on this, the greatest birthday, when I became a teenager. The song ended and I pushed Lucy back into my bedroom and played the record over and over while taking inventory of my Matchbox cars, which I stored in a white and red Campfire brand marshmallow tin. It was late and the house was quiet when I unplugged the record player and stashed the marshmallow tin under my bed.

"We l-l-listen too-morrow?" Lucy asked.

"Of course!" I replied. Music had become part of our language. By the time I turned fourteen, though, it was all Lucy could do to cough out "Amen" during church hymns.

SLEEPING IN THE car in the pouring rain made for a restless night. After I took a few tentative steps onto the muddy trail leading off beyond the fallen tree, an uneasy feeling came over me. I had become used to feeling safe walking through even the worst neighborhoods in New York City, and in South Dakota it was relatively easy to mitigate danger on the flat, treeless prairie by simply watching where you stepped. Walking into an unknown forest alone, though, felt like a different matter altogether. I couldn't see the horizon or the ground ahead of me. The mountains and trees and dense understory blocked me in, and like walking through a whiteout, I feared that what I couldn't see might kill me. I could cross paths with a hungry black bear. I could trip over a root and break my ankle. I could take a wrong turn and wander lost in the wilderness for days. I had walked only a few yards when I turned around and went back to write a note that I slipped under the car's windshield wiper. I explained that I had hiked past the roadblock looking for the Valley of the Giants, and I listed the date and time, my full name, and phone number at the winery, just in case.

After a few hours on the same track I was so deep in the forest that I couldn't tell in from out, north from south. It was so ill-maintained that it became little more than cross-country hiking on a faintly outlined footpath. Several times I stopped walking to make sure I was still on the trail, which was obstructed by thorny blackberries. Eventually it merged onto a logging road with primitive ruts left in the mud by bulldozers. Late in the morning, with my suit tattered and loafers caked with mud, I passed a recent clear-cut site. The brown slope contained a wide swath of logging rubble, a landscape ripped apart and littered with fallen branches and massive upturned root balls twice my height. Small green saplings peeked above jagged roots in a great field of massive stumps. The trail in that area became impassable and difficult to discern among the logging debris. A few scraggly, windblown Douglas firs clung to life. The lone trees that remained standing were scattered every half mile or so, and the brown space that separated them felt depressing and sinister, despite the expansive views. The logging companies had left the sentinel trees standing to reseed and replenish the landscape, but they seemed to cringe from their newfound exposure to sunlight. Misshapen branches skewered out at bizarre angles a hundred feet up towering trunks, like tiny arms on a *Tyrannosaurus rex*. They were reminders of the grandeur that once existed here and the enormity of what had been lost.

The sight of the wrecked landscape unsettled me. I imagined the bulldozers that tumbled over the hillsides and annihilated the forest ecosystem. I felt sad and angry about the deforestation, but those feelings included the complicated truth of my own contribution to it. By building a wooden canoe, I was, in a tiny way, complicit with the slaughter.

I clambered up the side of a massive tree stump. Its flat top, freshly cut, still oozed golden-brown sap. I counted the growth rings backward on the sawn wood, from the bark to the center pith. The years rolled away under my fingers—their droughts and floods, their lightning strikes and cold snaps, their seasons of abundant nitrogen and crippling pestilence, all written into the varying rings. When the countdown reached 1977, my birth year, I scratched an indentation with my thumbnail. I counted down two more years to Lucy's birthday, 1975, and thirty more years to my father's in 1945, and marked them both. I sat back on my heels and could not process the scale of it. These three little scratches were dwarfed by the five-foot girth of this trunk, a mere blip in the geological timescale. Could all our happiness and suffering really be captured by a few measly inches of tree rings?

I continued marking half-century increments until I reached the center and tallied them up. The tree was more than three hundred years old. I sat cross-legged on the stump and thanked it for the wood that it had sacrificed so I could build my canoe. I thanked it for all the other canoes, paddles, matchsticks, nets, poles, logs, and fence posts. I thanked it for the two-by-fours and construction beams that framed out entire cities, made millions in sawmill fortunes, and laid a bed of railroad tracks across the continent. I thanked it for the baskets and crates, the cradles and beds, the plywood and firewood, the cabinets and decks, the closets and hallways, and for every gift that Douglas fir had ever given to me, my father, and to the industrialized American economy. I knew how hard it was to grow back after being cut down.

It was hard to say whether I had incurred a moral debt to nature by using wood to build my canoe. In the vast scheme of the Pacific Coast Range, it didn't seem plausible. I had about three hundred board feet of lumber back home, a mere drop in the ocean compared to the fifty or sixty billion board feet logged in America each year. I had probably killed more trees printing my PhD dissertation than building a canoe. Still, sitting there, I vowed to put wood to use in a way that would assure its immortality, as a timeless thing of beauty, in a canoe to be treasured. I may not have possessed the skills to execute it perfectly, but I wanted my canoe to be a testament to the miraculous trees that were cut down to supply the wood, and a tribute to the tools I inherited. It felt right to acknowledge and honor the sacrifices made, by both the trees and my father.

At the edge of the logging area the flat-topped stumps and jagged upturned roots gave way to a virgin forest of Douglas firs, hemlocks, and pines. Their limbless straight trunks culminated in clustered umbrellas of dark green branches at least a hundred feet up. A few steps into the woods the air cooled and the light dimmed by half. A stream gurgled somewhere far off to my right. I felt the desperate sludge of loneliness. I wanted to know what it felt like to have a whole life and a whole family and a whole purpose, with everything working in unison and harmony like a forest ecosystem. I wanted to talk to someone, and not just anyone. I wanted to talk to Lucy and my father, to share this wilderness with them.

The trail dropped down the depth of a ten-story building into a thick salmonberry patch. For every large trunk that surrounded me, a hundred saplings sprouted from their detritus. Even the runts were big enough to dominate any forest back in New York. Suppose an artist had designed one of these massive trees, exactly as it stood. That single sculpture would be a revolutionary work of contemporary art.

Was this it? Did I really make it here? Hammered into the soft earth was a handmade sign with red letters brushed on a peeling white rectangle: Valley of the Giants.

The trail flattened out beside a stream. I took off my shoes and dipped my feet in the water—cold enough to numb all pain. Shallow, steel-colored water flowed over smooth rocks that harbored salmon in their downstream eddies. I walked barefoot, avoiding the slippery boulders turned jade by moss, feeling mud squish between my toes. I became intensely aware of my own body. The thick, humid air condensed on my skin. I made my way streamside for half an hour until I came to a place where a sublime and pungent smell stopped me in my tracks. I breathed in the lightness of spring rain, the musk of ripe earth and mushrooms, a sprig of pineapple, a dash of butterscotch and vanilla, and bitter notes of turpentine, resin, and minerals. I couldn't imagine the evolutionary purpose of such aromas, except that they completely washed my mind free of any other thoughts. There was no loneliness, career dysfunction, deforestation, smashed canoe, or dead father to worry about inside the old-growth forest. All sound had been muffled, as if absorbed by a sponge—nature's own noise-cancelling headphones.

Then I saw the thing that made me stop in my tracks. It was woven into the forest shadows ten yards ahead. I had missed seeing it before. It was too big to comprehend as a living organism. I was standing before a spectacular Douglas fir, a lone giant that by some miracle had remained unharmed for most of a millennium. The buttressed conifer disappeared in the mist above me, at least two hundred feet up. It must have been near the upper limits imposed by gravity, taking two days to transport water from its roots to millions of needles. Secondary roots trailed from its upper branches down to the ground. Mats of soil, teeming with ferns, had accumulated in the crotches where its lower branches met its trunk. The tree was an ecosystem unto itself. Burls as big as minivans bulged out from the bark. This single trunk wouldn't fit on a logging truck even if the Boise Cascade lumber company tried to harvest it. I measured its diameter by lying on the ground near the trunk, scratching the earth by my head, and inch-worming my way across. It was eighteen feet wide—three of me. Beneath this giant, my own body seemed freakishly small.

I turned away from the trunk with a profound sense that I was not alone here; I was in the presence of an inexplicable yet palpable life force. The scent of decomposing vegetation and humus reminded me of the cattail slough on Lewis and Clark Lake. Could it be possible that my father was there in the forest with me? No, probably not. Probably just wishful thinking.

The remains of a fallen giant sprawled beside the trail, the equivalent to a twenty-eight-story building lying flat on the ground. A placard said that at the time the tree fell in 1981, it was believed to be the largest Douglas fir on earth and its widest point near the base was thirty-six feet. A healthy forest, I knew, needed dead trees like this one. It was the natural cycle of things. Science has not even finished counting all the species that feast on the wood of dead trees. The fallen trunk held an incomprehensible, micro world. Sword ferns, liverworts, lichens, and leaves as small as sand grains covered every crease and seam of its bark. Even the moss carpet was itself as dense as a miniature forest, teeming with tiny insects and invertebrates. For all the struggles that had befallen this giant, I couldn't be certain it was dead. Quite the opposite, it had given life. Near the root ball at ground level, a grotto had opened in the rotten heartwood, big enough to park a car inside, and a new tree had sprouted from the open wound. Its branches drooped with sprays of needles and were crammed full of cones.

I touched the trunk, feeling the spongy moss on the bark, and an image flashed in my mind of the minister at the funeral talking about my father's love of nature and the outdoors. Even uprooted and lying flat on the ground, the tree continued living through the other beings that its death supported. It was not made to accept death as a dead-end street.

The rains came, pelting without mercy. I speed-walked back up the trail I had come down that morning, scrambled across the clear-cut hillside, and ran to my rental car. All the miles I had toiled through the forest were reversed in a few hours. Although I could have admired the giants forever, I was euphoric to make it back unscathed. I took the logging roads to the gravel roads to the paved roads to the highways, and found myself at the Portland airport, where I bought a change of clothes and caught the next flight to New York.

When I had settled into my seat, I felt a sudden clear focus and perspective. There was no time for anything inessential. My whole life depended on getting back to Long Island to repair the hammer damage I had inflicted on the strongback, and to roll up my sleeves and build the canoe.

Our flight path tracked east along the border between North and South Dakota, a place that people on both coasts viewed as unimportant, flyover country. I glanced around the business-class cabin. Most everyone else was asleep and unaware, of course, that we were thirty-five thousand feet above Faith.

The sky was clear enough to see hints of my old stomping grounds below, the place where I had experienced my first connections to the earth, to cycles of life rarely witnessed and poorly understood in the cities. It was flat yet uneven, a spring green moonscape, speckled with dots that represented ranch houses and barns at the intersections of endless, straight-shot gravel roads. I wondered if there was a gay boy in Faith right now, roping cattle, riding horses, and worrying about how he fit into a culture that was stacked against him. If there was, I hoped that he looked up from his horse, saw the contrails of my plane tracking across the continent, and knew that he was not alone.

CHAPTER 21

UNEXPECTED GUESTS

Back on Long Island, I fixed and realigned the busted strongback, and grappled with regret. My act of aggression was an attempt to play Old Testament God to the wood, to smite it into submission, but all I really accomplished was destroying my work and discarding one of my father's most cherished tools. His hammer was the casualty of my impatience.

I called Scott Roberts to order more lumber, and shared some of my troubles with shattered strips, broken windows, and bruised egos. I could barely hear him over the beeping forklifts in his warehouse.

"Wood is fickle, but its imperfections make it beautiful," he said. "Don't fight the grain, follow it."

I had to learn from the bitterness of my disaster. My instructor was not a book this time; instead, it was failure, an exacting teacher whose methods sometimes involved the cracking of wood and the spilling of blood. That sickening snap of the strongback breaking under the blunt force of a hammer was a plea: *don't do this until you understand me.* The wood taught me about itself, showing me what tasks to fine-tune.

I never knew the right amount of forward pressure to apply when cutting the strips until I knew how much was too much and walked back from the brink. To rip-saw canoe strips was to see myself shaped by the wood, even while I was shaping it.

There was plenty of feedback for me to observe, both in the sounds and appearance of the wood. When I scrutinized the boards closely, they were full of twisting grain, with varying intersections between the red heartwood and the white sapwood, which is the new growth at the edges of the boards that corresponded to the outer rings of the tree. The imperfections were not insurmountable, but they did have to be noted, respected, and worked around. The attack of the blade had to be configured to the wood grain that I was cutting, or I would be no different than the lumberjacks clear-cutting the Pacific Coast Range.

Scott had given me one other piece of practical advice: "Get yourself a nice set of featherboards!"

It was not surprising that I had overlooked any mention of featherboards during the fifteen seconds that I had casually skimmed the table saw instruction manual. After some frantic research and a trip to the hardware store, I learned about the gadget that could be the answer to a lot of my problems. Featherboards are flexible attachments for the table saw designed to hold wood securely against the blade to prevent the dangerous kickback that had sent wood flying through my patio doors. They also apply consistent pressure to the wood near the blade, allowing for more even thickness of cuts, which would correct the problem of my scarf joints popping and warping out from the forms. With my new featherboards bolted to the table saw, I vowed not to stop cutting until I had a pile of strips worthy of my canoe.

I marked up every board with a pencil to plot where usable strips could be sliced out from the grain to avoid the brittle knots. The methodical flow of wood across sawblade reminded me of the steady evaporation that lifts water up hundreds of feet into a giant Douglas fir trunk and causes it to bleed sap after being cut down. I slowly graduated to less severe kinds of mistakes. I learned how to move toward a danger point in the wood, then stop short before things got out of hand. I learned to respect the wood's strength and the saw's power, while pushing them both to their limits.

I had thought the problem all along was me, but that was only part of the story. My technique and choice of species were at fault; the canoe design was not. It may have been too much to ask that juniper be used for boatbuilding, no matter how much I had tried to persuade myself that I could change its temperament under the saw, and that any difficulties cutting it would be offset by its aesthetic beauty. It was clear that juniper would not change its nature for me just because I was building a boat, and I had no choice but to adapt. In the minutes after a strip snapped in the saw, when the fracture was fresh, I assessed what caused it, and derived a lesson from the break. It was often better to throw the strip in the burn pile and start fresh. That little strip was not meant to be today. It was okay, I'd do better next time.

Rip-sawing strips for days on end echoed the times I had watched my father use his knife to slice away the skin from a deer carcass, slip the feathers off a pheasant, or scrape the scales off a walleye. He separated lean cuts of muscle from the fat, skin, bones, and organs—a process not unlike sawing around knots to find the clearest cuts of wood. The solitary act of standing over the saw and moving my hands back and forth, guiding the wood with the stabilizing pressure of featherboards, was as close as I had ever been to the mind set of a lumberjack or a meat butcher. Successively, I harvested thin fillets from each board and placed them in neat stacks on the kitchen counter.

While my mind remained singularly focused, my body was screaming and resisting. The muscles of my back and shoulders were bound in tense knots. Through the course of each day, every so often I stopped to stretch and brace my hands against the wall or lie flat on the floor to shift the pressure off my back for a moment of relief, before staggering onward. Even as I was in a state of increasing physical agony, the work was fun in a strange, abstract way. This canoe may have been costing thousands of dollars and consuming thousands of hours, but what I got in return was the chance to switch off my brain and sink into the calming repetition of the work. It was hard to believe that not too long ago, the canoe had driven me into fits of uncontrollable rage.

I always faced the sliding patio doors while I sawed. In the warm late spring weather, I propped the doors open, blurring boundaries between inside and out, and giving the sense that nature had moved right into my house. I lived at the edge of a great suburban solitude, a densely developed yet surprisingly quiet world of beaches and dunes, salt marshes and scrub forests, rolling alongside mansions and saltbox shacks for miles. One early evening, a flash of red in the yard caught my eye: an American cardinal, a male in red plumage, plump like an apple. Slowly, without making any sudden movements that might startle him, I turned off the saw and reached for my father's binoculars. I focused on the red bird as it flushed insects from *Spartina* grass with its wings. The Dakota Sioux believed that seeing a red bird meant the soul of a lost loved one was visiting you. My father, being a Christian fundamentalist, refused to acknowledge anyone else's beliefs about symbolism, but he taught me that the appearance of a cardinal reminded us to keep the faith. He pointed to passages in the Bible, in Hebrews, where God used signs and wonders to comfort people with things like birds and rainbows. The cardinal turned and looked up at me, cocking its head to one side, studying me. Then it looked out toward the bay. I followed with the binoculars. A red-tailed hawk perched on the bulkhead, its feathers blowing in the warm southwesterly winds off the ocean. I focused the binoculars on the hawk, and in the distance over the bay, a large airplane floated past—a jumbo jet with four engines, thousands of feet above, so high that its fuselage reflected the setting sun. Dozens of flights from Europe passed by every day, making their long, slow descent into JFK. There was no noise, just the hawk's feathers blowing in the wind, the swells on the bay, and the plane crossing the sky. The cardinal flew off.

I wasn't kidding myself that this was a rare sighting of an endangered species that was sent by God to comfort me. Cardinals are common backyard birds in suburban New York. I willed the belief anyway, that it was some kind of sign. I wanted it to be true that this bird flew into my yard at just the right moment for me to notice him. I found it confusing that I could feel comforted by this bird—thinking of him in a nostalgic way as if he were the actual ghost of my father—while at the same time feeling a residual indignation toward him. My grief was complicated that way: a mix of anger, regret, and longing.

I rip-sawed strips deep into the night and the thought never crossed my mind that I ought to be tired. The winds died down overnight, and the next morning, the mirror surface of the bay was broken by a school of menhaden, a baitfish commonly called bunker. They swished their silver tails in unison at the surface until an osprey flew low overhead, threatening them with attack. The bunker disappeared into their own bunker, the depths of the sea. The red cardinal appeared, alternately flitting between the lawn and the outstretched limbs of the neighbor's juniper tree that hung over my house. All around me, the coastal ecosystem burst with new life, having been fed by the warm rains that fell every other day. A painted turtle with orange stripes on its shell munched spring dandelions on the lawn next to the bulkhead. I saw one like that as a kid and my father picked it up to show me its brightly colored underbelly. It had been sitting on the edge of a cattle stock pond on the ranch one day after a thunderstorm, in that moment of stillness when the prairie held its breath and you could hear a cow shift a hoof in the mud.

The linebacker of a deliveryman from Roberts Plywood showed up with my lumber replenishment and I asked him to leave it in the driveway so I could carry it into the house myself, piece by piece. So far, the canoe had not received any visitors to speak of and I was not about to let that change. The deliveryman probably would have been thrilled to see my canoe, but I didn't want him or anyone else to see my house in this condition. I understood that feeling sometimes, growing up—that something was good enough for me but would be too embarrassing to share with others.

Mom called from South Dakota later that afternoon to announce that she was coming to visit in June. That's when I realized what an intensely personal space my house had become, how easily it could fail to live up to outsiders' expectations, and how crushed I would be if it fell short. Mom's voice snapped me out of my wood-sawing and nature-watching spells and plunged me into an angst-filled state of mind. I almost managed to talk her out of it at first.

"Where would you like to go while you're here?" I asked. "What about Maine? We could drive up the coast, eat lobsters, and take in the sights."

"That would be fun," Mom said. "But what I really want is to spend time with you, to see where you live, and see this canoe I've heard so much about."

She wanted to come all the way from South Dakota to see my canoe? Perhaps I had talked about it a little too much. The canoe probably sounded more impressive than it really was in person. I should not have raised her expectations.

After Mom's phone call, I organized stacks of lumber, swept up sawdust, and paced around nervously. I still hadn't told her about how I rid myself of all my belongings before taking on this obsessive project. I kept things mellow in our conversations so she wouldn't know the true extent of my sadness. With my father gone, I wanted her to be free to grieve on her own without the added burden of worrying about me.

I shuddered at the thought of Mom seeing my house in this condition. There was not a spare bath towel to be found, nor any pots and pans. The guest rooms and living areas were barren, aside from the canoe, one chair, a mountain of lumber, and the workbench with its profusion of tools spilling off the edge. I would have to buy a few things to make the place look enough like a home. The clock was ticking, and I had to get back to sawing so I could then redecorate the house.

By early June, I finished sawing all the lumber, a process that left every exposed surface inside the house covered in sawdust. My back and shoulders were so stiff and sore that I could hardly move. I was making myself lunch when there was a knock at the door. I wasn't expecting anyone, and Mom wasn't due to arrive for another week. I nervously rinsed and dried my hands thinking that whoever it was, they couldn't come inside to see this mess. The second round of knocking was louder and faster—insistent. Caper barked and scratched at the back door. I crossed through the living room and den, shuffling around tools and offcuts, and glanced at myself in the mirror. I was not a pretty sight, wearing my canvas workpants and an old Cornell T-shirt, both with ground-in black walnut sawdust that looked like coffee stains. Sawdust clung to my skin and caked under my fingernails, and my socks and work boots had filled with enough sawdust to compost half the yard. I took off my safety goggles, revealing ovals of white around my eyes in stark contrast to the brown and red sawdust that coated my bald head. I was carrying my father's knife, which I had used to cut an apple for lunch, and the dust mask swung awkwardly from my neck. I was a mess, but it was not my appearance that worried me as much as the rest of the house.

When I opened the door, I was relieved that it was my friend Dave, whom I had not seen since our lunch in Brooklyn the day my father died. He wore large black designer sunglasses and a pink cashmere sweater wrapped around his shoulders with effortless style. A Coach overnight bag, with two-toned leather and monogrammed initials swung from the crook of his right elbow while he shook Nicorette tablets into his mouth.

"Ohmigod, hi! Are you really a boatbuilder now?" he asked in mock disbelief.

"Well, I guess so," I said, groaning.

"That's so butch of you, so . . . *substantial*."

"You look so . . . *citified*."

"You look like a raccoon," he said, gesturing in a circle to my eyes.

And just like that, our old rapport resumed where it had left off six months ago, almost as if no time had passed at all. I had first met Dave two years before, when we sat next to each other at a gay men's book club that met weekly in the organizer's SoHo loft.

"Hi, I'm Dave," he had said while reaching out to shake my hand. Then, without any sense of irony or consideration of the fact that we were total strangers, he added, "You look super uncomfortable and sad."

While everyone else sat in chairs around the room, I reclined flat on my back in the middle of the circle, with my rear end elevated off the floor by a donut-shaped pillow. I had had surgery a few days before to remove a benign growth from my colon, and by doctor's orders I could only stand up or lie down, but never sit. As I lay there staring at the ceiling, the room was washed out in faded shape-shifting pastels, a side effect of the Percocet I took at lunch.

The moderator asked each man to share something personal with the group as an ice breaker. Dave told us how he was a swimmer at UCLA in his youth and had missed the cutoff for the 1984 U.S. Olympic team in the breaststroke by one-tenth of a second. He still had the tight, muscular swimmer's build and broad shoulders of a man half his age. More recently, in his psychotherapy practice, he had taught a group of people with disabilities in the Bronx how to sing and dance like cast members in *Wicked*, the musical.

"Everyone wheeled up to give me flowers," he said. "And then they sang 'because I knew you, I have been changed for good.'" Dave choked up and the book club host handed him a tissue. I was next and was not in any condition to follow up on his riveting performance.

"Hi, my name is Trent. Let's see, I left my partner this spring, and I haven't seen or spoken to my father in years. Apparently he has cancer. Oh—also, you may be wondering why I'm lying on the floor. I can't sit because I have a surgical wound on my ass. It's nice to meet you all."

The laughter broke the ice, and over time the book club became a surrogate for the gay men's therapy that I desperately needed. Dave was much more talkative than I was, emotionally cooler, more artsy and queer, and opinionated. He voiced strong views about every topic imaginable and had a successful career teaching the use of theater in psychotherapy at New York University. He had become dear to me, and his timing now was perfect. He could help me navigate through the fog that had descended in advance of Mom's visit.

I led him into the house, past the canoe, to the patio doors. Caper bounded alongside. Dave looked around with his mouth ajar. He set down his bag but then picked it back up right away and wiped sawdust off the bottom. He wrinkled his nose and sniffed the air while unwrapping the sweater from his shoulders.

"Isn't Caper supposed to be white?" he said, stooping to pet the dog. Caper leaned into him, revealing a white patch on his flank where Dave's hand had wiped off the sawdust. "You need a bath, don't you, puppy?"

He walked over to the kitchen counter, and in one sweep of his arm pushed a pile of junk into the garbage can: wood scraps, Chinese takeout containers, empty wine bottles.

"If I didn't know you were a businessman, I'd think there was a meth head squatting here," he said.

He opened the refrigerator and gasped. There were only two things inside: a jar of dill pickles and a moldy cheese rind. He waved the cheese in my face, pointing at the half circle of bite marks embedded in it.

"Is this what you call dinner nowadays?" he said, throwing the cheese in the trash.

I slumped down in the chair and sawdust rose from the cushion in puffs. I coughed and covered my mouth with my T-shirt.

"I'm just trying to live my life here, getting by each day," I said, folding my arms across my chest. "You haven't even mentioned the canoe."

"Well, what do you want to tell me about that?"

I stuttered at first before spitting out all manner of canoe history and construction methods.

"It's amazing what you can do with a few simple tools," Dave said.

"Planking the canoe right-side up would be extremely uncomfortable, so it's actually planked upside down on the forms," I explained. "That way, I can stand upright to work at eye level. The only drawback of planking it upside down is that after I've created this thing, I have to flip it back over in the end."

"That sounds complicated in such a cramped space."

"Let me just build it first!" I said, while simultaneously figuring out the logistics in my head. With a lift and a tip, I picked up one end of the strongback to demonstrate how light it was, and the canoe shifted more than I expected. I set it back down before it toppled over; the floor trembled. "That wasn't supposed to happen."

"When did you learn the names of all the woods and how do you know all this boat stuff?" Dave asked. "Where did this even come from?"

The question puzzled me. I couldn't remember either. It surprised me, too, suddenly, to realize how much I had learned.

"I don't know, I did tons of research. It just kind of happened over the last six months," I said. "Also, lots of trial and error. I messed up pretty bad."

"You didn't take a woodworking class somewhere?"

"No, it just fell into place, sort of. I mean, I'm all alone out here on the ocean, what else was I supposed to do?"

"Look, I think your aloneness is beautiful," he said. "You're free and powerful. You run this high-caliber winery. You're crafting this elegant wooden beauty here—well, I mean you've started, but it doesn't look like much. Anyway, as I was saying, you're gorgeous and young, and you get to watch the sunrise over the ocean with a puppy at your side. I'm happy about you."

I remembered Dave's reply so precisely because of how much it meant to me at the time. He had just the right way of reading my mood and cheering me up when I needed it.

While Dave was complimenting my singularity, my mind drifted to Mom's impending visit, and how I needed to make my house a presentable place for her. The space was my workshop first and my home second, but that wasn't going to make her feel very welcome.

"Mom's coming in a few days. I suppose I should clean a bit."

"I'll help you fix the place up."

Dave spent the day helping me clean and took me shopping to buy simple furnishings. I finally threw away my sleeping bag. There was an empty patch of garden soil just outside the patio doors where we planted basil, cherry tomatoes, and zinnias. I bought plants that were fully grown and could go in the ground right away, adding an instant pop of color.

Instant was the key word inside the house, too. I picked out some reliable furniture, a sofa and rug, two beds, sheets, and towels. I hadn't wanted these things back in my life after I had purged so much, but now I had a problem—an empty house—that these things were designed to solve. We arranged the furniture, not bothering too much with how things looked, just concentrating on meeting some basic human needs so Mom would feel comfortable. There was a moment when I hesitated, worried that I was resorting to my old habit of doing things impulsively. The difference this time was that I didn't buy a designer couch to impress anyone or to make the cover of *Coastal Living* magazine. For all I cared, the furniture could be taken to the dump after Mom left. I didn't put the same stock in these things anymore. They were just stand-ins. I wasn't sure all my hasty improvements had done much good, but my house was livable now, and the real authentic piece of craftsmanship, my canoe, stood tall in the middle of the room.

The stage was set for Mom's arrival, thanks to Dave. All my anxiety about letting people see my canoe seemed a little silly now. I didn't need to feel self-conscious even if the canoe was a sea monster the size and shape of my grief. It was also the thing that gave me a purpose. What's fun about an easy project, anyway? No, I was growing to love my canoe for its failures and triumphs, for giving me an outlet to sweat, get dirty, fail, and try again. Like the wood, like me, like Caper, like most everything in my life, the canoe was flawed and temperamental, but mine all the same.

Before Dave left for the city, he asked if he could host his fiftieth birthday party at my house in September. He had grown up in suburban Los Angeles, and like me had been estranged from his parents. A posh lobster bake on the beach would be a sure sign that he had made it in New York—just as sure as my renting an apartment with a view of the Empire State Building had been to me, before I left all that behind. I agreed to host his birthday party here, but as soon as the words left my mouth, I regretted saying them.

"Does that mean I have to finish my canoe before then?"

"No, it can be just like this."

"Alright, I'll see you in September, and until then, I'll be here eating pickles and cheese rinds, alone."

"Chin up. Whatever ails you, you'll always find a balm because you're a searcher."

"I don't know. Do you think I'm doing the right thing, or is all of this too crazy?"

Dave faced me, put both his hands on my shoulders, and looked me in the eyes. He spoke with the calm, controlled tone I remembered from the day my father died.

"Most people just sit still. That's the paradox, though, right? To be quiet and slow down so you can move forward? I'm shit at that. In the meantime, you have your canoe."

CHAPTER 22

OH, MOM

All this worrying about the house had distracted me from the significance of reuniting with Mom. What would she say about the way my New York life had turned out—would it measure up in her eyes? She had always been the counterbalance to my stoic father, but how would that dynamic exist without him? Fortunately, I had Caper as my go-between. He was funnier and sweeter and cuddlier than I was, and I knew that, at the very least, Mom would be pleased I had raised such a good dog.

She came through the arrival chute at LaGuardia. I smiled when I saw her. Drivers stood in a ring at the mouth of the passenger exit, holding up names on pieces of paper. At sixty-eight years old with graying hair and a petite five-foot-two frame, she was a beacon of unruffled Midwestern fortitude amid the frantic travelers who whisked past her in dark suits, talking on cell phones. She wore one of her many quilted vests, which she'd hand-stitched to resemble a field of sunflowers.

"Welcome to New York!" I said while hugging her. She and my aunt had visited me on Long Island twelve years ago, a little while after I came out of the closet. My father stayed back in South Dakota. That trip led me to believe Mom harbored an unspoken conflict between her religious convictions and her maternal instincts. It all felt like ancient history now

"I got a foil-lined purse so pickpockets can't scan my credit cards," she said. "It's a problem in New York, you know, I saw it on *NCIS*."

I laughed, having forgotten that people from the plains viewed New York City as an ominous place with danger lurking around every corner. Her singsong accent stood out, with its extended *o*'s and *a*'s made famous by that *Fargo* movie. When I moved to New York for grad school, I scrubbed my own accent clean so people wouldn't suspect where I was from.

On the drive home, Mom and I made small talk but avoided any topics related to my father. We mused why the iPhone has an emoji for red wine and champagne, but not one for white wine. She boasted about my second cousin twice removed, who took his lovely girlfriend to prom and got a Division I college football scholarship, and her neighbor's son, who starred with his wife on a National Geographic channel TV show about backcountry veterinarians, and her quilt club friend's son, who was a guitar maker in Nashville and married a wealthy heiress from Myrtle Beach.

"They say country stars buy his guitars," she whispered, like she was passing along a secret that wasn't hers to share. "Maybe even Garth Brooks."

To most people, such details would seem like polite conversation, but it gave me a pit in my stomach whenever she bragged about the accomplishments and heterosexual relationships of her friends' sons. The stories were part of a familiar pattern and didn't come with comparable praise for my own life achievements. I should have been used to this by now, but it still made me feel invisible. Did she brag to her friends about me in that way? Maybe it was too complicated to explain to the ladies at quilt club that I made wine in New York and had male sexual partners. I also speculated that an alien spaceship had kidnapped my mom and replaced her with this chatty, modern version. She remarked how she enjoyed a cocktail while watching *The Late Show with Stephen Colbert*.

"Your father wasn't a fan of gin or Colbert or politics," she said with a hint of disdain.

There was the surface-level shock I felt while absorbing the facts that she was drinking liquor and watching late-night TV, but her subtext cut deeper than that. My whole life she had always called him Dad, and even when she phoned with news of his death, she had simply said, "Dad died." This was the first time I heard her use the words *your father* in a sentence. It felt intentional and distancing, like what she really meant to say was, "the man formerly known as *Dad* will henceforth be called *your father*, because he is dead and not my husband anymore."

"I might vote for Hillary in 2016, for no other reason than it'd be nice to have a woman in charge," she said. "*Your father* definitely wouldn't approve."

"Whoa, that's huge, Mom."

She shrugged her shoulders. "Times change."

SUZANNE PRESZLER, MY mom, grew up in the Wittmayer family in the more prosperous eastern half of South Dakota. They weren't well-off but probably seemed rich compared to the struggling farmers all around. Whereas my father was a second-generation American from a hardscrabble Ukrainian family, Mom was a thirteenth generation descendant of Sir John Howe, an English knight who landed in Plymouth, Massachusetts, in 1642. They met at South Dakota State University when she was in the Chi Omega sorority and he was on the rodeo team. After college graduation in 1968, they got married, and my father won the Northwest Ranch Cowboy Association rodeo championship in team roping. When he shipped off to Vietnam in 1969, Mom moved into a condo in Minneapolis to work an administrative job at St. Paul General Insurance. At a family gathering once, I overheard someone call her the original Mary Tyler Moore, but I didn't understand at the time how progressive this nickname was.

When my father came home from the war, they moved across the state to the ranch. Mom converted from being a so-called liberal Methodist, which allowed women to be ministers, to my father's conservative sect of Lutheranism, which didn't permit women to speak in chapel. I never asked how she felt giving up her big-city job to follow my father to the middle of nowhere.

On the ranch, Mom's culinary ambition was on full display every day. She kept old family recipes organized in photo albums. My favorite was seven words long and written in my grandmother's cursive:

~ Peach Pie ~
Just like apple pie, but less sugar.

This four-by-six recipe card encapsulated my family life: if you didn't know how to do something, you had better figure it out yourself.

I traced many familiar sensory memories of childhood back to Mom. At the chicken coop, holding a warm egg for the first time, and in the kitchen, her hands holding mine while I cracked it into a frying pan. After my father butchered a chicken, I shook the cutlets in a brown paper bag of flour, salt, and pepper, handing it off to Mom, who dropped the pieces into a deep fryer while Hank Williams sang on the record player about rodeos and hard work. At the granary, her shrill *come-kitty-kitty* as we walked together from the house carrying metal buckets of chicken offal, three dozen farm cats sprinting toward us. At the ponderosa pine next to the swing set, where she pulled cactus spines out of my bare feet. At the rhubarb patch, Mom gripping my shoulders firmly as we stood back twenty yards while the dog chased away rattlesnakes. At the big tractor shed, the smell of engine oil and hay, and Mom delivering fried chicken and sun tea at break time, her hair in curlers. At the door of the house, Mom rushing me inside and dunking my hand in a bowl of ice water after the bull slammed the trailer door on my hand.

I used to follow Mom out to her father's machine shop to watch Grandpa Richard tinker around. He was a gearhead and had enlisted in the army during World War II as a tank mechanic who greased the bottoms of the tanks so they could drive through water without engine failure. He owned a large Chrysler dealership and the pasture next to my grandparents' house was overrun by several dozen cars taken off the lot to be harvested for spare parts. I wasn't allowed to walk through the car graveyard on account of the rattlesnakes that hid among the endless mass of tangled electrical wire, abandoned transmissions, serrated tailpipes, broken windshields, and rusted hubcaps. Knowing Mom's side of the family like I do today, I can see that I was shaped by an inheritance from both sides. I came from a family with a common language in baking, building, canning, crafting, farming, making, and tinkering.

WE WERE BACK on the North Fork from LaGuardia by sunset. I had almost forgotten how nervous I was about my canoe, but I remembered as we pulled up to my house and walked inside. There it was, unchanged from this morning. I had hoped it would have built itself while I was gone. When Mom walked in, I expected her to say something along the lines of *Looks like you've done a lot of work here* or *Wow, look at all the progress you've made.* Instead she saw the canoe and said, "For Pete's sake, it's *huge!* You'd better not be spending all your money on this. You need to *save* your money because you *never know.*"

I mumbled that I only wished she could have come a few months later, when everything would be done, and the canoe would be ready to launch in the bay.

"My goal is to finish by the anniversary of—" I stopped myself short of saying "Dad's death." The words got stuck in my throat. It was June, and the anniversary was six months away. I couldn't believe it had been so long.

"Here, I brought you something from home. I felt so bad that the branding iron got lost," she said, handing me a gift-wrapped box.

I was surprised and curious what could be as important as the branding iron, or at least important enough to justify hauling it across the country in her carry-on bag. It was solid and heavy. I tore open the wrapping paper and discovered my old Campfire marshmallow tin. At first, I didn't allow myself to believe what it was. How could they have saved it all these years? I was amazed that Mom still had it, and both touched and surprised that she gave it back. I pried open the lid of this twenty-five-year time capsule and out spilled my Matchbox toy car collection, the ones I used to race around my bedroom with Lucy and the Muppets while listening to 45-rpm records. I picked them out, piece by piece, and turned them over in my hands. Besides the Matchbox cars, there was the yellow dump truck that I used to move small piles of sand around the brick patio, before I smashed it to bits. There was the petrified wood I found while fishing with my father on the banks of Black Horse Creek. There was a bronze letter opener that Grandpa Richard carried back with him after World War II. The things inside this can recorded events that they had witnessed, even things I couldn't remember or wasn't alive to see. How silly that these trinkets still mattered to me, that their shapes were all so familiar.

"Thank you for saving them, Mom."

She smiled and nodded. "Good memories."

We left the tin next to my father's toolbox and went out for dinner at a local café. Mom studied the menu for a few minutes and then set it down. "I'll just have whatever you're having."

"Get what you want," I said.

She folded her hands in her lap and looked around the dining room. "I don't know, you decide. It all looks delicious."

I was exasperated with her indecision. I really wanted her to order something that would excite her. I ordered for us anyway: fresh local sea scallops. The waiter brought over a bottle of my winery's own chardonnay and opened it tableside. Mom watched him use the corkscrew with intense focus.

"I tried to open a bottle the other day and I couldn't," she said. "Your father always did that for me, so I've been buying bottles with screwcaps."

It hadn't occurred to me that she relied on him so much that now some simple everyday tasks were made more difficult by his absence.

I told her that I would take her to the winery the next day and show her how to use a corkscrew.

While we waited for the food to come, I worked up the nerve to ask questions that I had been mulling over. I wanted to reach a different place in our relationship, where we could talk more openly.

"So, is there anything you want to talk about—I mean, about Dad?" I said.

"What's there to talk about?"

"Basically, everything that hasn't been talked about for years."

"No use rehashing it now. What's done is done."

I paused to take a bite of food and steeled myself for what I was about to say.

"I want to know if you have any regrets, because I do, and I want to share them."

She took a long sip of wine before replying with surprising candor.

"I wish your father and I had travelled more, seen more places. I suppose it was a money thing, but he worked so hard to save money and now what good does it do?"

"You could travel now, though! Make up for lost time?"

"Travelling alone isn't the same. You know that."

"Sorry. You're right."

"What about you?"

"I regret that I wasn't there in the moment Dad died. I wasn't there when Lucinda died, either. I don't know what it feels like to really show up for someone."

"You can't beat yourself up like that. It's not your fault. You had to go live your life in New York. You couldn't just sit around waiting, that wouldn't be practical, would it? Everything was in God's hands."

"I know I did what I had to do, but God's never been great about answering my prayers. All this would've been easier if he had."

"You don't know that, you *can't* know that," Mom said, growing indignant. "For all you know, maybe God did answer your prayers, but in ways you can't see yet. These things have to be revealed in due time." Her voice cracked and trailed off.

I shrugged my shoulders and poured another glass of wine—and then I wanted desperately to walk back from this conversation. I was in way over my head and I didn't want to upset her, so I tried changing topics abruptly. I swirled the wine in my glass, described its aromas, and summarized the weather patterns of that vintage. While I droned on about the wine, it seemed as if Mom wasn't listening. She stared off into the room, her eyes unfocused, like she was mulling something over. She interrupted my nervous wine babbling.

"I've been a wife and mother, always the caretaker, and it's hard to flip a switch, you know?" she said. "I want to take care of myself for a change."

She reached into her pickpocket-proof foil-lined purse and took out a wad of used tissues to dab the tears from her eyes. I watched the delicate way that she lifted her bifocals to do this. There was a bruise underneath the pads where her thick glasses had rested on her nose her whole life. The skin on the backs of her hands was loose and fragile, her nails painted a soft pink. She had curled and coiffed her hair and put on makeup and coral-colored lipstick. There also, shining amid all her fragility, catching the light, was the small diamond engagement ring that my father gave her in 1968.

In that moment, I saw Mom as a person, and my heart broke for her. I couldn't imagine sleeping next to the same partner for forty years and then one day waking up alone. I had been so focused on my own grief that I hadn't spent enough time considering how hard it was for her to lose her daughter and husband. I didn't even ask her about it in a direct way. I had just assumed that Mom was Mom, and she would get by like she always had. I began to see her in a different light, like some of her old Mary Tyler Moore spirit was fighting to break through after years of burden. I wanted to tell her that I loved her and would fight for her happiness like my own. Despite feeling all those things inside, when the moment presented itself, like usual, I froze. I rested my hand on hers and took a breath to speak.

"Oh, Mom," was all I could manage to say.

CHAPTER 23

LOST AND FOUND

The next day was one of those coastal Long Island simulations of heaven: seventy-five degrees and clear, the air smelling like briny seawater and grapevine blossoms. I took Mom shopping in the Hamptons. We walked into store after store where one glance at a tag exposed a showroom full of outrageous markups. Mom would pick up a bowl or vase off a shelf, turn it over to read the price, raise her eyebrows, and set it back down. I took her out for tuna Niçoise salads at Sant Ambroeus, where paparazzi staked out the front door, waiting for celebrities. By early afternoon, she had a weary, deer-in-the-headlights look on her face and I sensed that we needed to retreat to the more bucolic environs of the North Fork.

"I suppose if you worked in a cubicle in some skyscraper, then yeah, the Hamptons would seem like heaven," she said. "But they pay all that money for houses that are so close together!"

Being from the wide-open plains, it was easy for me to be cynical about the place where I lived. When I had first moved to the area to run the winery, I thought it was nothing more than a playground for Wall Street moguls to party on the beach. I didn't understand why this place was mythologized and exalted. It didn't seem all that special or natural compared to the *real* wilderness out west. As I lived here longer, though, I saw how the quiet beaches were a respite from the noise and congestion of the city. It was, after all, the place I retreated to for solitude after my divorce. I grew fascinated by the deeper history of the North Fork and how it came to exist. Going back to study the ancient geological past, I was surprised to learn about a commonality between the two places I had called home. South Dakota and Long Island were both shaped by the same glacier during the most recent ice age. The Laurentide Ice Sheet was so massive that its westernmost lobe flattened the Dakotas and its easternmost lobe piled up rock debris, creating the forked tip of Long Island, plus Cape Cod, Nantucket, and Martha's Vineyard. The ice took two million years to form, inching its way across the continent little by little, then melting quickly, in about ten thousand years. I strode each day across an ancient glacier's abandoned stage set, and I felt an aura of archaeological grandeur by living here—a sense that forces greater than me were at play.

I gave Mom a driving tour of the North Fork while Caper stuck his head out the back-seat window, his tongue wagging in the wind. We visited a rocky beach overlooking Long Island Sound. The sky was so clear that we could see the cliff face on the mainland, seventeen miles across the sound, where the glacier had gouged out part of Connecticut.

"The rocks the glacier left behind are still here, hidden underwater," I explained to Mom as I knelt and picked up a handful of beach stones. I let them pass through my fingers. The pressure from the glacier had rounded off any rough edges and tumbled their surfaces smooth.

At the winery, I parked around back to avoid the Ferraris and chauffeur-driven black Escalades crowding the tasting room parking lot. When we stepped into the vineyard, we entered my quiet domain among vines with taproots growing thirty feet down into the rocky, loamy soils. The vineyard rows stretched to the horizon, where they ended in a mixed hardwood forest. The gnarly old grapevine trunks were twisted and contorted from years of scarring, yet by some miracle, green shoots emerged from the bark and stretched into the warm June sun. There was a line I sometimes told customers when I gave tours of the vineyard and I repeated it to Mom: "We romance grapevines among the glacier's ruins."

I unhooked Caper's collar and off he went down a row of thirty-five-year-old chardonnay vines. The vineyard and the beach were the only two places where I could ever let him run free. He would be fine if no wild animals showed up. As we strolled and Caper bounded ahead of us, I explained the life cycle of grapevines to Mom. They needed three or four years to establish root systems before we allowed them to set fruit, and the grapes they produced from age five to fifteen would have vibrant, candy-like colors and aromas. They had their most productive years in their twenties and thirties, when the grapes exhibited their deepest colors and most complex aromas. At some point around age forty, the vines would begin a long, slow decline under the corrosive forces of disease and weather. Grape clusters would not grow as big and vibrant as they had in the vine's youth, and harvest yields would decline. The ancient Greeks believed that the oldest vines, between sixty and ninety years of age, produced so few clusters that the ones remaining were the most flavorful of any vines. The plants invested all their remaining resources in the last clusters of their lives.

"They sound a lot like people," Mom said.

Near the intersection between vineyard and forest, an osprey flew overhead, wobbling and erratic, making a high-pitched whistle. The bird looked like it might crash into the vineyard but righted itself and flew away carrying a bunker that it must have plucked from the sea only moments before. The fish flapped its silver tail, blood dripping out of holes where the osprey's talons had pierced its skin. Caper scampered ahead sniffing the ground until we reached the wildflower meadow that trailed down a gentle slope to the woods. I pointed out a pawpaw tree to Mom, which produces odd-shaped fruits that reek of bananas.

"It's the only tropical fruit native to the United States and nobody knows about it!" I said, feeling relaxed in my element showing Mom the natural world of this place that I called home.

Just then, with the recoil speed of a rifle, a deer shot up from the wildflower meadow. I snapped my head in its direction, then at Caper, off the leash, standing still. The muscles under his tight skin flexed. He growled. We witnessed a chain of events that happened so fast it was as though they unfolded in a series of high-speed freeze-frame photos. The deer bounded across the meadow. Caper streaked behind it in a white blur. The deer darted into the safe margin of the forest. I screamed Caper's name, urging him to come back, but he followed the deer. He ducked his head low and disappeared into the underbrush. I ran to the last place I saw him. There was no deer and no Caper, just a black hole on the edge of the forest that led into a thicket of poison ivy. Underneath the low-hanging branches of a white oak tree, Mom and I called his name. There was the distant rustling of leaves and breaking of sticks somewhere off to my right. I stood, unnerved, and held my breath. The rustling stopped and there was no sound at all that resembled my smush-faced, heavy-breathing, floppy-eared Caper. There was only the hushed flutter of oak leaves.

"What's on the other side of these trees?" Mom asked.

"The Long Island Railroad tracks, then a four-lane highway, some farms, and the beach we just went to," I said, shuddering at the thought of Caper rocketing out of the forest and crossing the highway.

"I wouldn't get too worked up about it," Mom said, sensing my panic. "Your father always said that farm dogs know their way home."

"That's the problem, Mom. Caper isn't a farm dog."

We waited by the woods an hour longer until the sun waned, and the sea breeze chilled the air. Caper didn't come back. Boxers were bred to hunt big game, so I couldn't fault him for doing the thing he was born to do. I blamed myself for letting him off the leash, but he was so happy when he was running free. I had friends who were cat people, and friends who had never owned dogs who could not understand when I tried explaining Caper's deep importance to me. Over the past year of my personal turmoil, Caper was the only constant. He was by my side through all of it, helping me absorb the body blows. He held together my life. The world that I imagined without him in it was a world that dissolved into nothing.

I called the police to report him missing while Mom and I retraced our steps through the vineyard. My vision was cloudy. I thought I might get sick. I stared straight ahead, trying to project an aura of strength. If I made eye contact with Mom, I might have lost it. My subjective sense of time shifted. Every passing second slowed as we walked back to the winery without Caper. By the time we arrived at my car, the winery was closed and the weekenders with the fancy cars had all left. We had been out walking for a few hours. It felt like years.

I dropped off Mom at home and drove around the neighborhood with my windows down, calling Caper's name. I must have knocked on fifty doors. I went to the police station and showed the officers the photo of my father holding Caper in his lap, sitting in his recliner watching football on Thanksgiving. I imagined Caper shivering in the woods somewhere if he was still alive. Driving around scanning for any signs of my beloved pup, all I could think about was the specter of seeing his bloodied and broken body lying by the side of the road. Sometime around midnight, I called off the search and, reluctantly, drove home.

Mom was ensconced on the couch with a glass of wine, stitching away on her quilt. She seemed relatively composed about the whole thing—at least on the outside—accustomed as she was to the ranch life of animals coming and going, living, and dying. I convinced myself that Caper could have been found alive, taken in by some family and given a new name, like George, and he would never again hear his own name. The people would think they lucked out finding such a cute puppy, not realizing that their George once sat on my father's lap before he died.

I slumped down on the couch. "Mom, what if he's dead?"

"Honey, stop right now. Don't even think about that."

"But these things happen every day. People's dogs get run over. Plus, he needs his heart meds!"

"Well, for Pete's sake, if you're not going to calm down then you should at least keep yourself busy."

She pushed a needle into the fabric draped across her lap, reached underneath, pulled the needle through, and poked it back to the top. She had pieced together various colors and shapes of fabric to create a realistic-looking house on her quilt. Under each window of the house, she had stitched a small white rectangle as a flower box, with tiny red geraniums growing. She had started sewing quilts around the time that Lucy died and continued making them to donate to injured active-duty military and veterans. She told me this quilt pattern was called Home Sweet Home, and that, back in the day, she would sew late into the night while my father worked in his shop. By two o'clock in the morning we both struggled to keep our eyes open.

Mom put her quilt down. "Tell me more about your canoe."

I hesitated. "Oh, I just cut up some wood pieces here, then glue them over there, and someday it'll be a boat."

"No, I mean really *show* me."

I felt an instant blast of anxiety. All my machinations over the past six months had been for my eyes only. Sure, Dave had seen my canoe, but he had no frame of reference for what constituted good woodworking, so my skills and, yes, my masculinity were not threatened. With Mom here, though, I had to exhibit some degree of proficiency with the tools she had steered into my hands, and that she had watched my father use before. I shrugged off the nerves, knowing that with Caper lost somewhere out in the world, I wouldn't be able to sleep anyway, so I gave in to her insistence.

I started by showing her the strips of juniper, basswood, and walnut that I had already cut. They ranged in length between four and six feet, and I explained how I would connect them end to end with tapered scarf joints to create twenty-foot-long strips.

"I tried this two months ago and it didn't end well," I said.

"What happened?" Mom asked, reclining on the comfy new couch cushions, sipping chardonnay.

"Let's just say, Dad's hammer is swimming with the fishes."

I described how I had recently read about bead-and-cove joinery, which involves shaping a rounded-off or half-circle edge on one strip to fit into a concave hollow edge on a different strip. By milling the strips with these precise edges using a router, one piece would fit on top of the other and they could be positioned at different angles. They would behave like the ball-and-socket joints in a knee or elbow.

Mom nodded with a look of tiredness. "Go on, then."

I stopped and put the wood down. "I really don't think I can do this right now. I'm worried sick."

"Suit yourself, but you might feel better if you just get one little thing done. That helped me when your father was sick. I'd look for anything to pass the time if it meant I could stop thinking about chemo for one blasted minute."

With Mom on the couch well out of range of the router table, I reluctantly pushed a few strips through it. The bead-and-cove edges came out remarkably clean. However, the high-pitched zinging sound of the router bit gave me the beginnings of a migraine. Everything happened in slow motion: scoring the scarf joints with my father's knife; cutting them with his hand saw; slathering the joints with wood glue; and clamping them together with jigs that resembled giant clothespins. My fingers grew numb. I sweated. The experience of trying to work on my canoe while Caper was lost and potentially dead magnified the failures of my life by a thousand. I was a pile of shit for letting my dog—who had a heart condition and an instinct to hunt deer—run free off his leash in a vineyard full of deer.

I stepped back to assess my work. I had managed to combine short, useless-looking fragments into a longer, stronger, and more purposeful juniper strip. I didn't really care. Without Caper, the canoe felt meaningless.

"See, Mom? I did it. I guess there's one scrap of good news today."

There was no response. The house was quiet.

"Right, Mom? Mom?"

She had fallen asleep on the couch. I hadn't even noticed. It was four in the morning. I turned off the lights and tiptoed to bed. I sank into my new mattress, a luxurious upgrade from the past six months spent in a sleeping bag. For the first time in as long as I could remember, I prayed. I asked God that Caper could be returned to me in one piece.

I woke up after two restless hours, tossing and turning in my sparsely appointed bedroom, and indicted by every banal thing: Caper's bed in the corner; his leash hanging from a hook on the wall; his bowl of uneaten food; and the bits of white stuffing on the floor, remnants of the squeaky toy he had disemboweled. I got ready for work without making a sound, to avoid waking up Mom. In the bathroom, I washed my face and screamed silently into a towel for a few fast breaths. This couldn't be the way it ended for Caper.

I drove to the winery, distraught. I couldn't stop speculating about the awful things that might have happened to him overnight. I stopped at the police station and there were no reported sightings of any stray dogs running loose in town. I arrived at the winery before any of the other employees. The grapevines were the same as those we hiked past the day before, but the vineyard felt different and somehow darker. I stood in the wedding pavilion with a view over the rolling green land, searching across the tops of the vines. My perch on the deck was essentially a ledge elevated above the vineyard. I had designed it that way on purpose, directing the architect a few years before to build the deck in such a way that our customers would believe they were floating. I saw nothing.

I reached into my pocket and pulled out a small, silver dog whistle that I had found in my father's toolbox. He used it when I was a kid to train his hunting dogs to retrieve game birds. I didn't think I would ever have a use for it, but here I was. I inhaled sharply, pressed it to my lips, and blew. I couldn't hear anything. I stared down the long rows of grapevines that trailed off toward the horizon in the soft morning haze. I waited for the events of the last day to reverse themselves, for Caper to come bounding out of the woods, wagging his nubby tail and telling me by the smile in his jowls that it had all been a practical joke. He didn't appear. God and Mother Nature and the universe, I had learned, were never, ever kidding. They would take whatever they wanted and would never give it back.

Dejected and feeling sick about the latest devastating heartbreak in my life, I walked around the side of the vineyard to the barn that housed my office. My keys jingled in my hand as I unlocked the front door. There—it happened lightning fast and in slow motion all at once—Caper sprinted up the sidewalk and flung himself onto me. I gasped and pitched backward under the force of his affection, barrel-chested and strong. He was all tongue and slobber and reeked of dead fish and deer shit, but I didn't care. We rolled around playfully on the sidewalk as waves of relief washed over me. I laughed and he licked my face. For once, I had unconditional love that didn't leave me. The universe had given him back. I felt less alone, and I finally knew how it felt to show up for someone when they needed me most.

Mom was outside pruning dead branches from the hydrangeas in the yard when I pulled up to the house with Caper. She saw us and raised her arms, letting out an elated "YES!" Caper grabbed one of the prunings in his mouth and ran circles around the hydrangea.

"I told you farm dogs always come home," she said, shaking her head and laughing.

Later that day, Mom helped me glue up my full-length sheerline strip on the canoe forms. I felt a new kind of closeness with her, one in which we both invested more in each other's lives. I steamed the wood by placing it inside a PVC plumbing pipe and pumping hot steam into one end of it with a teakettle. After an hour, the strip was as soft as a twenty-foot-long piece of cherry licorice. The brittle juniper that had given me fits for months finally relaxed. I balanced the strip amidships while Mom held one end. Without a hammer in my arsenal, I switched to my father's 1980s staple gun to quickly secure the strip against the twelve station forms. The sound of the stapler snapping the wood in place reverberated inside the house with a pleasing *thwunk*, like a drumbeat of happy music, telling me that this canoe was going to happen after all.

"I guess this means I'm a boatbuilder now," I said, feeling some pride. Mom started prepping dinner in the kitchen. I was happy as we slipped into our comfortable, familiar roles. As usual, she snuffed out my optimism with her dry, Midwestern pragmatism.

"It's a long way from floating," she said.

She was right, it was a long way off, and in boatbuilding there was no such thing as a consolation prize for effort. In the end, the builder is always judged on whether the boat floats or sinks, but that was a question I wouldn't be able to answer for several months. I would have liked to have stayed there and worked on my canoe a little longer, but the sun was going to set soon, and tonight was for celebrating Caper's return.

I set up a little table for us on the beach where we savored grilled rib eye and steamed lobsters. We toasted with cabernet franc and watched the sky grow dimmer. I told Mom the story about my trip to see the giant trees in Oregon, and how I planned to use the rest of my father's tools. Caper paced up and down the beach doing all his favorite things: chewing up driftwood, digging holes, and pouncing on crabs. For dessert, we toasted marshmallows over the grill and squeezed them between shortbread cookies.

The stars came out. Mom looked around in wonder. "I have to say, when we were on the ranch, I never would've guessed you'd end up in a place like this."

"Yeah, it's weird how things work out," I said, pouring her another glass. "And you *do* realize, of course, that Caper and I eat surf 'n' turf on the beach every night, right?"

"Oh, you joker, you do *not*," she said, reaching down to feed Caper a scrap of steak. "But I do know Caper is a farm dog at heart. Your father figured that out right away."

I looked around at the beach, at the bulkhead, and at Robins Island in the middle of the bay, with its twinkling light at the end of the fishing dock. "It's not a bad life, huh?"

"How do you remember to go to work every day? Living here is like being on a permanent vacation."

"Is that really how my life looks?" I laughed, and I tried to picture it through her eyes—my romantic, beach-bungalow-at-the-seashore life. "Why doesn't it look that way to me most of the time?"

"We have different perspectives, I guess," Mom said. "To me, this is all so fabulous."

I shrugged. It wasn't worth disputing or getting into how the last year I had spent in this house was the most difficult of my life. It was okay to enjoy the moment. I stretched out my bare feet in the sand. Small waves burbled against the beach stones worn smooth by the glacier. The two of us plus Caper sat in silence looking up at the stars. Occasionally, one of the stars streaked across the sky and disappeared. Where the sky dipped down to meet the ocean, the horizon was as dark as blue can get without being black. I thought about my father's funeral. When the officer on bended knee presented Mom with the folded American flag, I had reached over to touch her hand, and I felt the embroidery of the white stars against the navy blue background. Throughout childhood, I was moved by the blue at the far edge of the horizon, the color of solitude and freedom, the color of places where I could never go. Everything around us now was the same color, an exquisite and unknowable dark blue—except it did not feel like the color of someplace we were not, or the color of some faraway horizon we could never travel to. Instead, this color felt specific and immediate. My house, the bay, Robins Island, and Mom, Caper, and I were all enveloped inside the same, infinite blue. For the first time in a long while, I believed that my family was safe from destruction.

I didn't need to work up the courage to say the words that I never managed to squeeze out in the presence of my father.

"Love you, Mom."

The barbecue illuminated her face while she held a bamboo stick with a marshmallow over the glowing embers. "Love you lots, honey."

CHAPTER 24

LITTLE AND OFTEN MAKES MUCH

Strip-planking the canoe after Mom flew back to South Dakota went easier than I expected. With each strip that I steamed and stapled to the strongback, I accepted the two possible fates: either it would lie smooth against the forms and resemble something Gilpatrick would be proud of, or it would warp, twist, and crack to oblivion. Whatever the outcome, I had to accept it. I wanted the canoe to think it was all part of the plan. The wood, I had come to believe, could sense my insecurity.

After installing several more juniper strips on top of the sheerline, on both sides of the canoe, I decided to create a decorative racing stripe using two woods: white basswood and black walnut. The inspiration came from Mom and her deft touch with the various colors and textures of fabric in the quilt she was making when she visited. First, I ran short strips of basswood and walnut through the router to create the bead-and-cove edge. Then I scarf-joined them together into long strips, let the glue dry, and steamed them in my PVC pipe contraption until they were pliable. The steamer leaked water on the living room floor, which Caper lapped up as though it tasted better than the water in his bowl two feet away. I glued, clamped, unclamped, adjusted, and repositioned the strip numerous times before achieving a satisfactory fit. After stapling it into position, I checked that row frequently throughout the day, expecting to see warps, bumps, and cracks. Nothing went wrong! The woods fused seamlessly, as though they had grown from the same tree. My confidence soared.

Despite all my efforts in the aesthetics department, progress was slow, and I had only built up six inches of hull on a three-and-a-half-foot-wide canoe. At this point in late June, speed was more important than precision, as I still had in mind my goal of finishing in December. I spent an entire weekend on a rampage of scarfing, steaming, gluing, and stapling until I had laid up ten new strips. The staple gun filled my house with many *thwunks*, music to my ears. I didn't focus so much on the details; I had mastered those techniques once and didn't need to replicate the same level of exactitude every time.

It became apparent within minutes of finishing those ten strips that something was wrong. They fit flush in some places but in other places, they slowly separated, with sticky strings of half-dried glue stretching between them. The main difficulty was bending the wood around the bilge, or the point of maximum curvature between the horizontal bottom and the vertical sides of the boat. Rounding the bilge required the wood to undergo a compound bend with a high degree of torsion: from horizontal placement amidships, twisting to vertical placement at the bow and at the stern, with the added pinch inward as the hull tapered. Around the bilge, the strips lost contact with the forms and the hull ballooned out, giving it a bloated appearance that reminded me of the time Socks gorged himself sick in the oat bin.

Then there were the staples. I had used up whatever came loaded in my father's staple gun and bought more at the hardware store. I grabbed the first box off the shelf labeled "staples," but didn't check if they were stainless steel. Who knew there were so many kinds of metal alloys? I didn't think this would matter until black stains streaked down from the wood at each of the hundreds of staple holes. Apparently, the galvanized steel had reacted adversely with the natural aromatic oils in juniper. I would need stainless steel staples from now on. The combination of these two mishaps meant I had to reinstall the ten warped strips. My frustration grew as the days wore on. My motivation plummeted. I got tired just thinking about trudging home from the winery and then spending hours sweating over the bilge. Everything, it seemed, involved repetition, waiting, and failing, then deciding if I had the energy to try again.

My boat didn't resemble Gilpatrick's slender and graceful design. I was no nautical engineer, but it seemed that a hull with daylight visible through cracks would probably sink to the bottom of the bay. I could end up with a boat that didn't float, yet I would have no way of testing its seaworthiness until I had already wasted a year building it. My confidence as a boatbuilder—which I thought I had built up in a way that would sustain me through this project—was at risk of crashing down. When I was in the presence of the old-growth trees in Oregon, I had vowed to put the wood to use making an object of timeless beauty. My work so far embarrassed me, though, this enormous, excessive, ridiculous, bulging object in my living room.

I distracted myself with housekeeping: mopping floors, scrubbing bathrooms, folding laundry. Every hour or so, I stopped cleaning to lie on my back underneath the canoe, hoping it would have made a miraculous shift into position, and the gaps would have closed themselves. There would be no miracles. The canoe would remain a potbellied, overgrown freak of nature until I learned how to fix the mess I created.

I didn't know how to proceed. I flipped open *Building a Strip Canoe* and found not a single mention of what to do about strips separating from the bilge. How many strips would it take to build up the hull, anyway? I hadn't considered this question with any sort of quantitative clarity—a big oversight. I had been so excited to start building that I didn't stop to count the strips or think through the scope of the project. So I located a full-color photo of a finished canoe in the book and counted the number of strips with my finger ticking down the page.

Fourteen, fifteen . . .

Okay, not bad.

Thirty-two, thirty-three . . .

Hmm. I glanced at the calendar on the chalkboard wall.

Sixty-seven, sixty-eight . . .

Good Lord. My heart rate built slowly to an uncomfortable thump in the bottom of my throat.

Seventy-nine and eighty.

The canoe required eighty strips; I was in full panic mode. It was almost July and it seemed unlikely that I could finish on time. My entire project was in doubt.

AFTER SELLING THE family ranch, we moved four hundred miles east to a town called Yankton and lived in a handicap-accessible apartment in a low-income development called Green Hill—which was neither green nor on a hill. My father worked overtime as a welder and came home covered in soot. Lucy and I slept on the floor in sleeping bags next to my parents' bed, trying not to make sounds to disturb them. On good days, Lucy sat in her wheelchair and hummed along with Patsy Cline records. On bad days, she slept and moaned and choked on her pills. I started seventh grade in a three-story brick building that had an elevator with glowing electric buttons—the first elevator I had ever ridden. And some of my classmates, I noticed, were the first people I ever saw who had a skin color different than mine. I waited in a separate line to receive free school lunches, and do-gooders brought us fruit baskets around the holidays. After school, my father made me trudge through the soybean fields outside of town hunting pheasants to put food on the table.

A few months after middle school classes began, the principal called me into his office to inform me that my father had suffered a massive heart attack that required quadruple bypass surgery. I thought he was absolutely going to die, but he pulled through and was back to work six months later. The school's music teacher, Mrs. Lyons, heard about our plight and offered us her second home, so we left the apartment complex with the misleading name and moved into a house that had a dirt floor in the basement, three different colors of paint on the outside, and an aggressive dog chained to a tree in the backyard. I thought things were finally looking up.

Then one night, when I was a freshman in high school, Mom was driving Lucy home from the children's hospital and got sideswiped by a drunk driver who ran a stop sign. Mom's shoulder blade was shattered to bits, her organs were bruised, and her kneecap was torn off her leg. Lucy was immobilized with two broken arms and two broken legs. They needed hospitalization for weeks, if not months. I wanted more information from my father, some statement of their injuries or their odds of survival, even if it was just so I could turn the words over in my head and dissect them with reason as a way of coping. I wanted him to say *The cop thinks they'll be okay* or *I reckon it don't look good* so I could at least process the horrifying scene. Instead, all he said was this: "Trust in the Lord that He has a plan for all of us." There was no precedent or road map for navigating emotional experiences with him.

While my father worked constantly and nursed Mom and Lucy back to health, I begged him for my own car so I wouldn't have to rely on him for rides to school and my work shifts at the local burger joint. After the series of misfortunes my family had endured, though, I knew what the answer would be.

"We ain't got the money," he said. "You gotta earn it."

We drove to the social services building so I could sign up for the Rent-A-Kid program—a community service where people around town hired kids from poor families to do odd jobs, like yard work or babysitting. The hallways of the Rent-A-Kid office had the antiseptic vibe of a hospital. Beige linoleum stretched into the distance, framed by yellow oak doors and wall sconces with triangular green glass. The woman behind the desk wore round glasses on the end of her nose, and her gray hair was pulled up in a voluminous beehive bun secured with chopsticks. When we walked in, she was sifting aimlessly through a large pile of papers. We pulled up chairs to her desk at an angle, like we were going to be looking through documents together.

"I was hoping you'd have some work for my son here," my father said. "The name's Preszler." He removed his greasy black welding hat and smoothed the few thin wisps of hair across his sweaty head. The woman looked at me over her reading glasses and started filling out a form with my relevant information, address, and age. After a few minutes of scribbling, she picked up a sheet from her pile, gazed at it blankly, and let it drop.

"I'll send you to Old Man Anderson's place out by the stockyards," she said. "He's a cranky SOB but he pays on time, in cash."

"That's good, thank you, ma'am," my father said, standing up to leave.

"Here's the address," the woman said, handing me a piece of paper. "Should be a pretty straightforward job, I think."

My father dropped me off and left, as he was already late for work. Nobody was there. The street was bleak and airy, wide enough to handle the constant flow of eighteen wheelers that rumbled past on their way to Yankton's livestock auction. It was blazing hot, ninety degrees, not a cloud in the sky, and not a tree within sight to offer shade. I had nothing to eat or drink with me. The property had an old barn on it, but not a fancy barn with soaring gables, cupolas, stained-glass windows, and doors that slid open on iron wheels. No, this barn was made of bricks, with a flat roof and the long rectangular shape of a warehouse. There were no windows, and one door, which was locked. In the olden days, it was probably used to fatten hogs or slaughter chickens. Waist-high ragweed choked the property. I walked up to the short side of the barn and found a piece of paper on the ground by the locked door. It was wrapped around a small wire bristle brush with rubber bands. I unwrapped it and read the note: *Rent-A-Kid, Use this to scrape off all peeling paint*. There was no further instruction nor other tools provided.

I had never scraped paint before, much less white paint—probably leaded—that peeled from the rough surfaces of bricks and mortar. I started in tentatively at first, making long, indiscriminate swipes at the bricks at random places along the wall. Within seconds I had scuffed my knuckles against the bricks so hard that all four of them on my right hand were stripped of their skin. It was the most miserable day of work I had ever experienced. By the end, I was dehydrated and starving, with a deep sunburn and both hands crusted with bits of coagulated blood and white paint chips. My fingers cramped from gripping the brush. Worst of all, after that first day there seemed to be no tangible evidence of my excruciating work. I could barely tell the difference between the regular bricks and the ones I had brushed. There wasn't a before and after, just misery.

The next morning, I was a little jumpy at breakfast. It took me several tries to find the right words to convey to my father that I *was ready* to work hard, but not if it involved scraping paint off bricks. Eventually I settled on the pity-party approach and showed him my scabbed knuckles.

"I'm not going back there," I told him. "It's the most miserable work. I hate it."

"Yes, you is goin' back, and I don't wanna hear yer moping. Git in the car."

On the drive there, I dialed up the intensity a few notches. "Look at my hands! I can't even bend my fingers! I can't believe you're forcing me to do this!"

My father's eyes rolled against my pleading. "It's on you now, and you're right, it ain't gonna be easy. Nuthin' is. All yer belly aching would be easier for me to sympathize with if I didn't have bigger fish to fry."

"I know, I know," I said. My problems were nothing compared to Mom's and Lucy's, but they were still my problems. We arrived at the barn and my father stood in front of it with his thumbs wrapped around his belt buckle.

"It's a big barn, I'll give ya that much," he said, shaking his head. "Could take all summer."

"I'll never finish. I don't even know where to begin. And all I got was this measly wire brush."

"Alright, alright, I've had enough of yer ballyhoo," my father said, raising a hand to stop me. He surveyed the enormity of the barn and walked all the way around it through the ragweed. When he emerged from the other side, I folded my arms across my chest in exaggerated protest.

"Or," he said, looking at me with his hazel eyes magnified and unblinking as he grabbed the wire brush out of my hands, "we could figure out how yer gonna git the job done."

"Well," I said gingerly. I was heartened by the *we*, but this was an awful job and I needed to make sure he understood that. "*We* could, but it's impossible."

"We're not gonna let ourselves believe that," he said, walking up to the broad side of the barn. It towered over him. He knelt and brushed the paint off a single brick, then handed me the brush. "Git down here and do the next one."

I knelt beside him and brushed the peeling white paint off a general section of bricks, covering an area about two square feet.

"Stop right there, you're all over the place," he said, ripping the brush out of my hand. "That's not what I toldya. Just do ONE brick. Focus!"

"Okay, jeez, relax, Dad."

I brushed the paint off one brick, a tiny surface area, eight inches long and two inches wide. "See? At this rate, I'll be here forever!"

My father took the brush and scraped the next brick in the row, then he stopped. "See this here brush? It don't know it's gotta scrape the whole dang barn. It only knows this one brick."

"That's my point, there must be a million bricks!"

"Yer stubborn head's gittin' in the way," he said, taking off his square, wire-framed glasses and wiping sweat off his face with one long swipe of his shirtsleeve. "Don't you remember anything we talked about back on the ranch? Tell me something that happened all at once there, when the whole shebang blew up."

"Oh, I forgot," I said, hanging my head and fidgeting with the brush. "There's fire, and tornadoes, and lightning."

"That's right, they swoop in real quick and get out fast. You don't wanna be like that."

"Nope."

"And I reckon you remember something that happened real slow like?"

"Yah, the trees."

"Yup. How'd you s'pose that ponderosa got so tall? It sure as heck didn't decide one day to shoot up ten feet. All it knew was gittin' a little bigger every summer."

We took turns brushing the paint off bricks, tacking our way across the seemingly endless broad side of the barn. My father coaxed me onward, admonishing me to move slow and steady.

"First this brick, then that one," he said.

It worked, kind of. After a few minutes of brushing, I felt a sense of spiraling dizziness from what I swore was heatstroke. I left my father's side and stepped around to the north side of the barn for some relief in the cooling shade. Then I went back to the broad side of the barn and found my father had not missed a beat. He was still kneeling there and had finished scraping the first row, all the way across.

"You see that? We did the whole row one brick at a time," he said while squinting through the glare of hundred-degree sunshine. "Little and often makes much."

Looking at him standing before the barn, the tide of fatigue dragging at my own teenage body, I felt terrible. Though I resented my father most of the time or found him impossible to talk to, he had just turned fifty and gone through a string of hard times: the total loss of his ranch; his wife and daughter's near-fatal car accident; his daughter's permanent and worsening brain disease; his own heart attack; his Vietnam service; his once celebrated rodeo career long since forgotten. And yet he was still here, helping me scrape paint off this stupid barn. If I had any idea it would have hit me this hard, I would never have let things get this far and would have scraped the barn on my own without dragging him into my silly teenage summer job. It was too late, though. He was already here.

"Dad, you should just go," I blurted out, waving my arms, and pointing down the street. "I'll figure it out myself."

Apart from that one stroke of inspiration in the beginning, the work was tedious, excruciating, hot, and endless. I did a decent enough job scraping the paint, but not outstanding. No one was about to notice, or care, that I took a lot of naps on the cool, shady side of the barn. The lack of contact from the Rent-A-Kid agency didn't feel like a good sign, either, but most of the time I resigned myself to doing the work, brick by brick. Every now and then I had a moment—most often it was snapping awake from a nap in a cold sweat—when the full potential of the job came down on top of me. If I quit, and could not buy a car, any future college scholarships I might get would have to be forfeited and my dreams of studying botany would evaporate in the summer heat. My fate would be sealed. I would have to slink back home with my tail between my legs, and I wouldn't be able to hide my failure from my father.

Growing up the way I did, one thing was an absolute fact: having a car and a little gas money would give me freedom. So, I stuck with it, brick by brick, and by the end of that summer, I finished scraping Old Man Anderson's barn. The Rent-A-Kid lady paid me, and I marched into a used-car dealership and handed over $1,750 in cash for a silver 1988 Ford Fiesta with a hundred thousand miles, a wonky timing belt, an after-market sunroof that leaked, and a rust spot on the floor where I could watch the road beneath whiz by. All of that, but it was sweet because it was *mine* and I had earned it the hard way, the patient way, my father's way: little and often.

"I'll never scrape another brick the rest of my life," I told him when I got home from the car dealership.

He laughed. "I reckon you'll always have the scars on your knuckles to prove it."

HIS ADVICE CAME in handy later in life, too, when I was in grad school at Cornell University. My research consisted of individually numbering thousands of grape clusters on vines in the field, then clipping them off at intervals throughout the growing season and measuring their weights, sugars, and acids. It was mind-numbing, monotonous work that took four years to complete, and I could only get there one way: cluster by cluster. When I finally walked across the stage to accept my PhD, my father wasn't in the audience. I hadn't seen him in years. I dedicated my doctorate to him anyway. I would never have finished without knowing that little and often makes much.

That's what it would take to finish my canoe. The only logical path forward that I could discern involved removing the botched bilge strips and starting over from the sheerline. I would have to make eighty new strips, deliberately and steadily. I would steam and staple up one wood strip, make sure the fit was perfect, let the glue dry, get a good night's sleep, and add another the next day. First this strip, then that strip, nothing more, only discipline. My father would have approached the canoe in the same way. His advice had never let me down yet. Any time in my life that I repeated ordinary little tasks often, extraordinary things happened.

CHAPTER 25

SUMMER BY THE SEA

As row after row of strips fastened onto the canoe, it slowly changed shape in my house. Day by day, the steam warmed the wood. The cracks stayed away. The strips held firm, one each day. Each new strip brought with it a new challenge to fit it tightly onto the preceding one, which I addressed slowly through trial and error until I got it right. My fingers bore the brunt of it, and I sported a rotating patchwork of Band-Aids and gauze. I didn't so much mind my injured hands, though, or my constantly aching back; I was excited to see the canoe growing bigger right before my eyes. More of its graceful curves became visible and, for the first time, I got a real sense of its scope. By the end of July, I was in a deep rhythm.

My everyday life and boatbuilding merged into one and the same thing—two perfectly matched pieces of wood, glued and clamped together, without a sliver of daylight between them. The sawdust found its way into my flower beds as mulch, where the cherry tomatoes that I had planted before Mom's visit now flourished. Strips still broke or shattered now and then, and my foibles with the table saw produced lots of odd-sized offcuts, but I was happy to use them as kindling in the barbecue. I sought out artisanal butchers and experimented with grilling methods using different species of wood. I cooked old family recipes from notecards Mom shared for dishes like the beef Stroganoff and peach pie. My refrigerator was no longer the barren receptacle of moldy cheese rinds that Dave had discovered; now it overflowed with fresh kale from my own garden. In a way, using sawdust and offcuts to grow and cook my own food allowed me to eat my mistakes. In the process, I also regained an appreciation for life on the ranch, and the rewards of being more self-reliant in the world. I never anticipated that the canoe would renew my interest in cultivating the land, too, but seeing how Preszlers had always survived by building and growing things, I was living up to my last name.

At first, I hardly noticed the change of seasons. Summer on the beach looked pretty much like spring on the beach, with a few subtle differences. The days were longer, and the air was warmer. I took special notice of the sun shining just beyond the sand dunes, where the water was blue green instead of its usual gray. Redwing blackbirds called from the salt marsh, where they perched on impossibly thin blades of *Spartina* and filled the air with their unmistakable, crackling songs of summer.

On the first Saturday in August, with the patio doors open and my fortieth strip glued up, I took a break. The loss of the branding iron was weighing on me. I decided to make my own, a canoe logo of sorts. I brainstormed on my chalkboard wall, making notes of what my logo could look like until I ran out of space. The wall around the taxidermied duck was mostly full of my ramblings already, and the opposite wall still had the large monthly calendar on it. I found my father's red handkerchief in his toolbox, wet it under the faucet, and started erasing the calendar, smearing then wiping away the months of January and February. I swallowed the tight, emotional lump in my throat that didn't have any business being there. A calendar is just a calendar, surely nothing to get choked up about. As I carried on erasing the months of March, April, and May, the same feeling arose. I stepped outside to get some fresh air and collect myself. Caper was asleep on the patio. Nothing else moved in the heat and humidity of August. A distant thunderstorm over the ocean appeared in the most mesmerizing, sweet-smelling sky. I shook off whatever I was feeling and moved quickly back to the wall to erase June and July. It was then that I understood the root of the sentimental twinge. Those painful months of my life were over. My tortured winter was long gone. The uncertainty of spring had faded away. The anger of early summer came and went before I even had time to realize it was leaving. And now here I was in August, charting my future.

I twirled the chalk between my fingers for a few moments and pondered what to write in the blank space where those months used to be. Any good branding iron had simple graphical lines or linear shapes to symbolize ownership and identity. I thought about the things that really mattered to me, things that I would want permanently branded to my canoe, and wrote down three words, at first.

Wood, which I represented with arrows pointing at each other, that could have been fishtails but symbolized a canoe floating on the surface of calm water that reflected its mirror image.

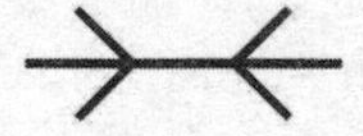

Water, which I represented with a raindrop.

Sky, symbolized by a circle, for the moon and the sun, which guided my days.

The symbols needed something to tie them all together. Caper woke up from his sun-drenched nap on the patio and ambled inside for a drink. A breeze wafted in, smelling of garden compost, fish, seawater, and the honeysuckle blossoms trailing over the bulkhead. That's when it dawned on me and I added one more word to the chalkboard: *Life*, which I represented with a nautical compass of the four corners of the earth—north, south, east, and west. I showed my chalk sketches to a graphic designer friend in Brooklyn and he turned them into a crisp logo, a copy of which I would send to a custom branding iron maker. If this one got lost in the mail, I could order another one. I had to laugh: all these months after I wrote a fake name on a credit form to buy lumber, Preszler Woodshop now had a logo.

I clicked send on the branding iron order and went down to the beach to go fishing. My father had left his old fishing reel inside his toolbox, and I had recently purchased a fiberglass rod to pair with it. I hadn't gone fishing in too many years to count. The day was hot, with clear skies and a perfect surf that erupted into gorgeous foamy waves onshore. The sun, which was about to set, cast a deep golden light over the beach. As I walked from my bulkhead steps toward Marratooka Point, light caught the spray coming off the waves, framing the beach shacks and sailboats in a misty halo. A young couple, a guy and a girl, silhouetted against the sunset, played Frisbee in the sand. When the Frisbee went in the water, the woman dove in after it, and the man gave chase until he caught up to her and grabbed her by the waist, swinging her around and kissing her until the waves receded and they walked out of the water holding hands. Theirs looked like the kind of romance that everybody wishes for, the kind that only happened in the movies or, apparently, right here on my beach.

As a gay man seeing their public affection, I had to perform a special equation in my mind. In my youth and the years that followed, it had always seemed to me that love between two men was a story that was never credibly told—or was told merely to elicit laughter or describe oppression. Public displays of affection were an institution to which I didn't belong. When I saw the young couple making out on the beach, I became agitated. If I wanted to feel any kind of empathetic response to their romance, I had to will myself into their experience with a selective, internal somersault.

I walked past the canoodling young lovers and cast my lure into the channel, where in the distance yachts glided along in slow, grand fashion. The engine of a yacht that had a helicopter landing pad on its upper deck was especially loud. Two yachts maneuvered around Robins Island at once, and the engine from one echoed the other's in a gas-guzzling duet. The sights and sounds were reminiscent of summertime fishing excursions I had taken with my father on the banks of the Missouri River.

WE SAT ON five-gallon chicken feed buckets holding fishing rods that my father had made from flexible bamboo. I pulled in junk fish—skipjacks—and my father cut them up for bait, but after a few hours of not catching anything, we started thinking of those skipjacks as supper. Save for the time I reeled in a ten-pound pike, we never caught trophies, but watched with silent envy as boaters pulled up walleye from deeper water with reels that looked heavy and expensive. Some of the boats were made of smooth white fiberglass and had chrome trim and built-in rod holders that made me embarrassed about our flimsy homemade rods. The people on those boats laughed and drank and listened to music in their colorful shorts and deck shoes, while I stood onshore wearing scuffed cowboy boots caked with manure. It didn't matter how many times we saw the fancy boats offshore, my father always had to comment, muttering something under his breath about *them danged kids, laughing like they ain't got a care in the world and driving boats that cost as much as houses.*

That was about the only time he ever spoke when we went fishing. He would more likely be rubbing his forehead and wincing, seeming to disappear right in front of me. His eyes would grow wide and vacant as he stared off into the distance, and I followed his gaze but never saw what preoccupied him. Like the realms of the young lovers and the boat owners, my father's internal thoughts were a world to which I didn't and couldn't belong.

I GOT SKUNKED. I reeled in my line and walked home.

Three weeks passed in the sweltering dog days of August, when it seemed like the entire population of New York City transplanted itself to my sleepy little hamlet. I experienced the annual invasion of carefree, beach-going, wine-sipping tourists. They double-parked on my street, partied on my beach, and stood in long lines at my favorite restaurants. They got sunburned and drank rosé while I went on with my daily routine: scuttling to work and back home again, folding laundry, paying bills, and sawing, routing, scarfing, clamping, steaming, bending, gluing, and stapling one more strip to my canoe.

On the last day of August, I asked my neighbor, John, to help me carry my canoe outside to the patio. After two months and sixty-five strips completed, the sight of my canoe hulking there in broad daylight looked as if Gilpatrick's model boat from *Building a Strip Canoe* had broken away from the page and floated up to my house by mistake.

I glued up a strip under natural light for the first time, feeling the hot summer sun on my face like I was one of those sentinel trees in Oregon suddenly exposed after a clear cut. Tourists waved to me and gave a thumbs-up as they passed by on Jet Skis and bobbed around on inflatable unicorns. I smiled indulgently when an older gentleman pulled onto the beach in his rowing skiff and sauntered up to have a look at my canoe. I would be lying if I said I didn't enjoy the tourists' approving glances. I was proud, showing off a little bit, being one of the admittedly esoteric artisans—along with cheesemongers, painters, potters, sculptors, shepherds, and winemakers—who made the North Fork different from the Hamptons.

After two days, I moved the canoe back indoors to protect it from the elements, and there it was in my living room: a boat hull that practically looked like a canoe, with only a thin, football-shaped hole in the bottom yet to be filled with strips. The once jumbled and chaotic mountain of scrap lumber had been organized and arranged into the kind of canoe that I had dreamed about.

With Labor Day came the unofficial end of the tourist season. I dressed in baby blue seersucker for a special occasion: hosting a VIP luncheon at the winery for the owner, Michael, and one of his friends, a prominent art museum director in Manhattan. I was ready for the event by the time it rolled around, despite the fact that my hands were wrapped in bandages.

On the way to work that day, I found myself stopped in a long line of cars creeping forward at a snail's pace. Without anything else to do but wait in the traffic, I looked out the driver's-door window, to the other side of the road, where the beach gave way to a verdant green pasture. It was there that I saw something so unbelievable that it could not have been real. Perhaps it was a mirage or a hallucination. I pulled over and parked on the shoulder of the road, then got out to have a closer look. Caper followed close behind, timidly. When I reached the fence, I saw that it was not a mirage. Standing there, a mile from my house, twenty yards from the beach, next to a merlot vineyard, was a small herd of cows. They lumbered up to the fence and let me pet them. It had been twenty-seven years since I sank my fingers into the curly hair between Herman's eyes. I felt their raspy, sandpaper pink tongues on my hands. Caper loved the cows, too, sniffing their noses through the fence. Their fur was pure white, the defining trait of Charolais cattle, the same breed that my father raised. How could they exist here on Long Island, of all places? As it turned out, a local farmer had acquired the cattle to supply fresh beef to the finest restaurants in New York City. His was the only working cattle ranch on Long Island. Having a vineyard and a herd of Charolais on my left and an ocean full of boats to my right felt like more than a coincidence. It was a reassuring sign from God that I was living in the right place at the right time.

Michael and his art museum friend, Glenn, arrived at the winery in a gleaming chauffeured Maserati. I led the winery tour; we sat for lunch; the staff presented overflowing platters of lobster salad, heirloom tomatoes, and filet mignon arranged just so alongside antique milk pails bursting with white peonies and ripe persimmons. If there was a question about the wines, I answered in detail, but otherwise I didn't say much. My rightful place was to be the wine geek hovering in the background, not to be the center of attention.

When I had asked Michael permission to return to Cornell for my PhD seven years before, in 2008, I thought for sure he would fire me on the spot.

"You can go to Cornell, but you can't quit running the winery," he said. "You're like a son to me."

He was a shrewd Hollywood negotiator and no doubt trying to appeal to my inner child, the farm kid who was estranged from his father. Michael graciously agreed to continue paying my salary while I studied. Receiving his blessing to go to grad school while still working for him was one of the greatest professional affirmations I could have hoped for. In Michael, I had found a surrogate father who didn't care that I was gay.

I had thought Michael was the most erudite and sophisticated man I had ever met, until I met Glenn. I hung on his every word. I poured a straw-colored wine called viognier into glasses with large bowls and laser-cut rims and explained that the grape was native to the Rhône valley in France. Michael and Glenn nodded approvingly and lifted the glasses to their noses to inhale the aromas of apricot and honey. Then I did something I had done a million times before: I dabbed the droplet of wine from the neck of the bottle onto a linen napkin that draped over my left forearm. As I did this, Michael's eyes grew wide. He locked his focus onto my beat-up and calloused hands.

"What happened? Did you get in a fight or something?" Michael asked.

I hadn't told my boss that I was building a canoe. I didn't think he would care. He was off in Hollywood producing films, and since my father died, I had tried to keep all my conversations with Michael limited to my job.

"Um, I cut myself . . . woodworking?" I said, squirming, trying in vain to cover my hand with the wine-stained linen napkin.

"I don't follow," Michael said, setting his glass down.

"Maybe I forgot to mention it," I said in a casual tone. "I'm building a canoe . . . in my house . . . with my father's tools."

Michael appeared to blanch, but before he could say anything, Glenn interjected.

"*A canoe?* I love canoes!" he said, almost jumping out of his chair with excitement. "I keep some at my lake house in Vermont. I'm Canadian, so canoeing is in my blood."

Michael seemed a bit bewildered how the conversation had veered from Picasso to canoes.

"That's fantastic!" Michael said. "Good thing Trent's hands look like they went through a meat grinder or we might not have discovered this bit of intrigue."

I laughed nervously and asked Glenn to tell me more about his canoes. He described a beat-up old aluminum canoe for everyday use and a canvas-on-frame canoe, and his pride and joy, a collection of ancient canoes. He owned priceless birch bark canoes that would merit space in a natural history museum had he not collected them himself.

"What's yours like?" he asked.

"It's a wood strip model, nothing too special, really not a big deal," I said, trying to downplay the whole thing while growing uncomfortable being the center of attention.

"I have an idea, let's look at it after lunch," Michael suggested with enthusiasm, picking up on the fact that our guest—one of the most influential people in the entire art world—would rather spend his vacation day away from the museum talking about anything besides art.

Michael set one foot in my house and wood shavings crunched underneath the trim soles of his suede loafers. I didn't have air-conditioning, relying instead on the regular sea breezes to cool down the house, but on this day the breeze had stalled. The living room was a sweltering sauna. Nervous about their opinions, I rambled to fill the silence, explaining how I had been working on the canoe all year. "I'm learning on the fly," I said, self-conscious. Whereas at the winery I was confident in the domain where Michael paid for my horticultural business expertise, that confidence wilted once I had welcomed him into my home.

In addition to his career running Hollywood's most successful independent film studio, Michael had been a contemporary art collector since the 1960s and was a trustee of Glenn's museum. Their world was one in which I was most definitely an outsider. They occupied a rarefied realm of Oscar-winning, culture-defining icons, to which I never expected having any special access. Their opinions didn't just matter to me; they mattered, period. Yet, the gap between our worlds shrank in the presence of my canoe. We became just three regular guys having a conversation. The canoe was the great equalizer.

I pointed out a few places where the joints didn't align, and where small cracks and screw holes would have to be filled with putty. Michael hushed me.

"First rule of being an artist: never draw attention to your mistakes," he said. "They're only mistakes to you."

Glenn chimed in that he had visited thousands of studios that were in disarray, with paint cans everywhere and unfinished works piled in the corner. "Art can be messy," he said.

"I'm just trying to get by each day without cutting my fingers off," I said.

"Regardless, your canoe is beautiful," Michael said. "I've never seen anything like it. Totally unique and totally tremendous. Glenn, what do you think?"

"It's a beautiful canoe, but it's too functional to be art, and too new to be an artifact."

"It's not functional yet," I said. "I still have to close up this gap, sand it smooth, and fiberglass it. If it floats, then sure, you could say it's functional."

Michael took a defensive stance. "I think it absolutely *is* Art, with a capital *A*. Look at it!" He folded his arms across his chest, and I saw that he, too, had sweated through his shirt in my sauna of a house.

"I beg to differ," Glenn said. "This can't be Art if it's going to float."

"I could show this canoe to any collector and tell them it's Art, and they'd believe me," Michael said. "If you don't think it's Art, then what is it?"

"It's a boat," Glenn said matter-of-factly. "If it floats then it's a boat and nothing more."

"An artful and artistic boat that he made with his father's tools," Michael said, growing impatient.

"Correct, but it's a boat, not a sculpture," Glenn said.

"Look, I never said I was trying to be an artist, and I'm not even a real boatbuilder. I've never done this before."

Glenn and Michael discussed how I could turn the canoe into contemporary art if I wanted to. I could cut a hole in the bottom and paint the word *fuck* across it in giant letters. Then I could sink it on purpose in the bay, while recording videos of the sinking to display at a Chelsea gallery.

"Now *that* would be Art," Glenn said.

I nodded my head while processing what they had just described. If that was the definition of Art, then I didn't want to be an artist. I'd had enough of tearing things down and questioning everything; I wanted to build and create now.

Michael scanned the chalkboard wall behind me, silently absorbing the things I had written there for inspiration. The word *F L O A T I N G* still loomed large around the wood duck. He ran his hand across the surface of my workbench and rubbed sawdust between his fingers.

"That's my father's toolbox," I said, opening the scratched lid so Michael could look inside. He leaned over awkwardly and gave me a hug with a firm pat on the back.

"Let's set aside this Art-versus-Craft debate for now," Michael said. "I think we can all agree that your father would be proud."

CHAPTER 26

THE WHISKEY PLANK

Feeling exhausted from the summer-long grind, I hoped that the hardest parts of the canoe build were over. The finishing touches should be all downhill from here—at least, that's what I was counting on. My friend Dave would be hosting his fiftieth birthday party at my house in late September, two weeks away, and I didn't have time for any more hiccups.

A few days after Michael's visit, my logo branding iron arrived in the mail. I unwrapped it, and it was perfect. The branding head was heavy, with a long metal handle and insulated wooden grip. I held it up to the canoe in different spots around the hull, imagining how well the five-inch brand would fit in scale on the twenty-foot boat. As I was figuring out where to position the brand, I ran my fingers along the joints and seams and inspected the canoe from top to bottom. Michael said artists shouldn't point out their mistakes, but alone in my house, being my own worst critic, I couldn't resist making notes about the areas that needed fine-tuning.

The part of the hull that remained to be filled was an oblong opening in the bottom, shaped like a narrow, seven-foot-long football running along the keel, or centerline. I stood on a chair and stared at the football from above, finding it mildly alarming but not entirely surprising that the gap in the hull was asymmetrical. The starboard side appeared a little bit straighter than the port side, which bulged out a couple inches. Without realizing it, throughout the summer I had glued up strips unevenly and given the canoe a crooked backbone. Perhaps one of the sheerline strips was off by a smidge and the error carried through to all the strips I laid on top of it. An untrained eye might not have noticed the subtle inconsistency, but I couldn't stop staring at it. Nonetheless, at this point I had to keep going.

The crucial piece of wood needed to fill the last gap in the hull was called the whiskey plank. There was a long-standing boatbuilder's tradition that involved drinking a shot of whiskey after installing the last plank in a boat. My palate was obviously trained for wine and I didn't have much fondness for whiskey, but according to legend, bad luck would befall me if I ignored this tradition. I didn't need any more misfortunes, so I treated the whiskey plank with a healthy dose of gravitas.

The tradition was not intended to celebrate finishing the boat but was more of a cautionary toast to mark a major milestone in the building process. Closing the hull wouldn't make the canoe seaworthy, either, until I fiberglassed it. Nonetheless, I allowed myself to feel an inkling of relief that I had come so far to face a stage in the process that I never thought I would see.

Mom told me that my father had brought precious few things with him to Vietnam: his rodeo rope, his camera, and his whiskey flask. Sadly, the camera had been pawned off years ago, and although he left me his rodeo rope in the toolbox, I had no idea how I could ever use it for the canoe build. The rope sat on my workbench collecting sawdust and getting in the way. But his silver whiskey flask was right there in the toolbox, waiting for its moment in the spotlight. I picked it up and dusted it off. I could only imagine the things it might have witnessed during his lifetime. I was sure he never could have imagined how I would end up using it, either. It was familiar from our hunting trips, and from one of the photos that Mom and I discovered in the same shoe box as his Bronze Star. In that photo, he was reclining in an army green pup tent next to a box of Triscuit crackers and a liter of Canadian Club that Mom had mailed to Vietnam from Minneapolis. I immediately went to the liquor store and bought a bottle to fill his flask.

It was time to cut the whiskey plank. As it turned out, there was a technique for determining the precise shape that a plank needs to be to fit a gap in a traditionally built wooden boat. It's called spiling and refers to measuring and drawing a shape onto a board. One of the marvels of boatbuilding was how all these miraculous strips fit together, given that a single piece of wood could be wide amidships, tapered from end to end, and then narrowed further to a point and curved up at the bow into a lazy *S*. As a boatbuilding technique, spiling intimidated me right from the start. I had read dozens of descriptions about it, but never fully grasped the process until I tried doing it myself.

To start, I drew a straight line down the middle of a juniper board, bisecting it. Then, at each station on the canoe, I measured the inner opening width of the football gap using my father's steel calipers, which were engraved with a date stamp of 1880 beside the name of a blacksmith in Boston. By transferring the station widths from the hull gap to the juniper board, I learned that counting and numbers weren't necessary for spiling. All that mattered were the relative widths transferred by the calipers: a unitless measurement. When I had the widths for all twelve stations marked on the board, I coped around the shape with a jigsaw and out popped this oblong football of juniper. The image of the whiskey plank mirrored the gap in the hull—a case of something imagined becoming something real. In a half day's work, it was ready to be fitted and installed. Spiling, up until this stage, was not particularly fun, but I liked having finished it.

I positioned the whiskey plank over the hull and found that it was decidedly larger than the gap it was supposed to fill. That was by design; I cut it too big on purpose. Inside the traced line was the bulk of the whiskey plank, which had to be preserved at all cost. Outside of the line was a fungible area where I could file down the edges to fit. I swiped the farrier rasp across the edges of the plank, then set the plank back on the hull to see how well it fit. Removing microns of wood at a time and continuously going back and forth between my bench and the canoe—little and often—seemed like the most logical way to tackle this stage. After an entire day spent measuring, rasping, and staring, I had what I thought was the best-fitting whiskey plank I could hope for. The fit was so snug, though, that when I jammed the plank into the hull with a ham-fisted thump, it snapped into three pieces, fell through the strongback, and clanked on the floor. A whole day's work was wasted.

I cut another whiskey plank, gaining confidence in the technique as I worked. I placed this one over the hull gap and made little adjustments with the rasp, too. Numerous subtle indentations marred the edges of the football gap in places where the bead-and-cove router bit had sunk too deep. The irregular peaks and valleys of the shape were vexing. I turned off all the lights in the house except for the bulbs directly over the canoe, then I slid underneath it like a mechanic changing the oil in a car. From that vantage point, I could see light in the places where the whiskey plank and hull did not exactly match. They were close, but it was an imperfect puzzle piece. I kept working through dinner, continuously filing down the plank to size, but each time I removed wood from one side, the opposite side was thrown out of whack. I removed too much material and the fit got worse every time I tried to make it better. I gave up for the night and smashed the whiskey plank over my knee—more for the burn pile.

One time as a kid, when my parents were out of the house, I gave Lucy a haircut. My first attempt at her bangs was uneven, the right side shorter than the left. So I cut more off the left. Then the right looked too long, so I trimmed that, too. Then the left was off, and on I went. Before I knew it, Lucy was crying, and her bangs resembled a bristle brush.

"You gotta learn to leave well enough alone," my father had said to me at the time, gently removing the scissors from my grip.

That was my experience filing down the whiskey plank to fit the football gap.

Tired and frustrated, I slumped into bed. There had to be a way of matching the differentials between the plank and the asymmetrical gap. I needed to replicate a mirror image topography of the hull—but how? Everything hinged on my ability to close the final gap. I had come too far to let this step of the process trip me up.

Stumped, and unable to sleep, I turned on the light and read a book called *The Genius of Japanese Carpentry: Secrets of an Ancient Craft*. It was an architectural classic that told the story of craftsmen who restored a twelve-hundred-year-old monastery. I flipped through the pages aimlessly, scanning the photos, admiring the impeccable craftsmanship on the ancient temples. Then my eyes locked on a passage about a topic that stopped me in my tracks: *wabi-sabi*.

It was a Japanese worldview that embraced the beauty found in imperfect objects, especially those that occurred in nature and had rough surfaces and asymmetrical forms. One could choose to go with nature, or against it, but the unavoidable truth was that everything trended toward chaos. Beyond the beauty of imperfect natural objects themselves, though, wabi-sabi also spoke to our ability as humans to accept and nurture imperfection in our daily lives. It was an aesthetic of impermanence, where cracks were highlighted, rather than hidden. Mistakes were accepted without shame.

My father had told me in the duck hunting blind years ago that sometimes my shots wouldn't hit their mark, but if I tried my best, that was good enough. As with cutting Lucy's bangs, I shouldn't let perfection become the enemy of good.

I closed the book and practically tripped over Caper as I ran into the living room in my pajamas. I switched on the overhead lights. There it was: my imperfect, mistake-riddled canoe with an asymmetric gap in the hull. Chunks of dried glue squeezed out between strips and dripped down the sides. Staple holes littered every surface like buckshot. Well, maybe I overstepped wabi-sabi a bit and veered into a new worldview of my own creation, which I dubbed *Really Ugly*. Either way, I didn't have to hide my mistakes or feel shame about them. Of course it was not going to be a pristine masterwork; it was the first thing I'd ever built. The old-growth Douglas firs in Oregon weren't perfect either, yet their scars made them enchanting.

Under the amber haze of my shop lights, sometime around midnight, I spiled one last whiskey plank. There were a few minor cracks and gaps around the edges, but I could fill them with putty and fiberglass later. For now, it was close enough to fine. I coated the edges with glue and slowly, carefully, shoehorned the seven-foot-long plank into the gap with the bronze letter opener that my grandfather carried home from World War II, and which Mom had given back to me in the Campfire marshmallow tin. The whiskey plank snapped into the hull with a most satisfying *thwunk*. I toweled away excess glue squeezing out from the seams.

I had finally solved the riddle of assembling the eighty strips.

Closing the hull didn't resolve the imperfections in my canoe, but rather embodied them through wabi-sabi. Considering this, the questions I had at my father's funeral about who he was in relation to me felt less urgent. We had both tried our best but fell short of being the man each of us hoped to be for the other. That was the essence of it, anyway. I had been churning over this internal struggle for so long that meaningful closure seemed out of reach.

I stood in the presence of a rough and unfinished form, hulking yet elegant, not yet smooth or seaworthy, but fully realized in its overall shape and dimensions. This was no ordinary boat, but the timeless silhouette of a grand Iroquoian canoe. Somehow over the past few months, building it had come to feel like a part of my normal life. The idea of finishing it scared me most now.

There had been months of panic of not knowing what to do or say, but a new feeling was at the heart of my life now: I was not a watcher anymore. I was now a doer, a maker, a builder, accountable to myself alone and to the things I brought into the world—however imperfect they were.

I lay on the floor underneath my canoe, propped up on a pillow with Caper beside me. One more time from the top, I read my father's Bronze Star declaration, letting myself go all the way to the end. It was sad and thrilling. I lifted the whiskey flask up toward the canoe in a toast—to Dad.

CHAPTER 27

A PARTY

My life was a cycle of the seasons. Dad had educated me in the rhythms of farming, driven by constant renewal. Calves were born, fattened up, and sent to slaughter, so that more calves could be born in their place. Wheat seeds were planted in bare earth, sprouted green, matured in amber waves, and tilled back into the soil after harvest. All I had known in my life were circles of perpetual change that when completed started me back at the beginning.

By the time I finished installing the whiskey plank, summer had waned, seeming to evaporate in its own heat. The days were still hot, but the nights had begun to cool, and the chilly hours after dark grew longer. A dry September made for ideal grape-ripening conditions. The throngs of city folks had departed the North Fork, and we locals were left to enjoy the last few weeks of summer in peace before the whirlwind of autumn harvest season began. Without really planning it this way, the natural progression of my canoe build paralleled the seasonal cycle of the vineyard. September would be a time of quiet preparation for a different kind of harvest bounty: paddling my own canoe.

The small yellow crowbar with a forked tip in Dad's toolbox was the perfect tool for the next step in the building process. I used it to pull out thousands of staples that held the strips to the mushroom-shaped forms. I had some moments of trepidation while removing the staples, concerned that the wood might suddenly pop out of shape after being freed from its shackles. The hull stayed true, though. Being pinned against the forms all summer long had bent the wood permanently—not unlike me, molding my daily life around the canoe's construction until it felt permanent and natural, and I could relax.

Based on Gilpatrick's instructions, I washed off the now staple free hull with a sponge and hot water. By some miracle of wood's elastic cellular structure, the hot water expanded the grain and swelled the staple holes shut until they were practically invisible. I breathed a sigh of relief watching the holes disappear. It was almost like the staples had never existed in the first place. Renewal was the work of time, or in this case, hot water.

Next, I smoothed the rough and jagged stems, or the ends of the canoe at the bow and at the stern, where all the strips came together in a point that would someday slice through the water. For this, I returned to a most trusted tool that I had grown confident using by now: Dad's farrier rasp. I swiped it diagonally across the stems and watched the unmistakable shape of a canoe materialize. The stems on each end looked like the shoreline of a crescent moon bay. I could hardly stop touching these elegant sculptural shapes that hadn't existed in the world until I created them.

The hull at this point was still composed of hundreds of scarcely visible facets at the seams between the strips. The unevenness had to be faired, or made perfectly rounded and smooth, using a hand plane and sandpaper. Dad's 1910 Stanley hand plane was a tool made for smoothing rough wood by applying force to the handle and driving an angled blade over the surface. The metal heel had a dark brown patina that was cold to the touch when I picked it up for the first time. I passed my finger over the blade gingerly, noting with appreciation that Dad had left it razor sharp for me. I expected as much; I could not imagine him letting a dull blade leave his shop.

Suppose I had started smoothing the boat's eighty-square-foot surface area haphazardly, without taking a systematic approach. Not only would I have been overwhelmed by the task, I might have run the plane over some areas multiple times and removed more layers of wood than necessary. Likewise, I might have unwittingly missed some spots. To avoid such scenarios, I drew pencil lines across the hull, dividing it into ten smaller sections. I swiped the plane across them in a deliberate fashion: first this section, then that one. Once the pencil marks for a section had been shaved off, I moved on to the next, applying Dad's little-and-often approach to problem solving. It allowed me to tackle any complicated, multidimensional job without the old flashes of panic.

The first time I pushed the hand plane across the hull, the action resonated down to my core in an elusive but elemental way. Part of my pleasure came from the joy of mastering a new tool that I had never used before. There was a constant figuring and calculating as I worked, related to the geometric angles at which the plane sliced most cleanly and cleaved most evenly across the hull. The other part of the satisfaction was less academic but more sensuous. I liked the way the juniper, basswood, and walnut strips murmured with a low-pitched creak, as though they came alive the moment the blade touched them. I also liked how the wood resisted at first under the force of my hands, before letting go of paper-thin, translucent ribbons.

As I shaved off the facets with both hands on the plane, I felt the connection between what I was doing and what Dad had always done with his hands. My reward for this deliberate, unhurried work was a sudden and mysterious unfolding of beauty. Lovely and unpredictable patterns of color appeared on the planed surface: streaks of chocolate brown from the walnut and golden ivory from the basswood harmonized with the burnished red of juniper. The curled shavings perfumed the air with intoxicating, spicy-sweet aromas. Each pass revealed swirling wood grain arranged in concentric ovals like fingerprints, each one unique, and never before seen. The unheralded lumber that other boatbuilders had passed over was yielding something beautiful. I was hypnotized by the constant revelations the smoothing brought.

Although I would not say I had mastered the hand plane, by any means, I had used it well enough to accomplish the task within a few days. I crouched down at the freshly planed surface and ran my hands over it, feeling for hidden knots and irregularities. I studied the annual growth rings the blade had uncovered, reading subtle cues of shape, texture, and grain to speculate on how the trees had grown when they were alive. Looking back at the canoe through the patio doors from outside, I saw a slight waviness in the hull. The imperfection was elusive. I could stare right at it and only after several minutes of looking did it materialize. If I changed my angle of sight or stood farther away, the bumps disappeared. I couldn't keep shaving off more wood, though, and at some point, I had to stop before I frittered away all my hard work. Using Dad's red paisley handkerchief, I rubbed the steel sole of the plane with paste wax to prevent rust, and then I placed it carefully back inside the toolbox.

At a marine supply store down in the harbor, I bought a fairing board, a flexible piece of thin plywood reinforced with fiberglass. Velcro was set on one side to affix sandpaper. It was a deceptively simple tool designed to round off any remaining seam facets and smooth over any grooves left by the blade of the hand plane. To protect us from the dust, I slipped on my mask and locked Caper in the den. Within the first few strokes of the fairing board I became aware that the boat felt to be under subtle but continual tension caused by the unreleased compression of the strips—something akin to a drawn bow and arrow, waiting to be sprung. This gave the hull a kind of supple liveliness as it bounced from the pressure of the sanding, as though unseen life forces lent the boat elasticity and resilience. It would bend, but not break.

A September heat wave rolled into the North Fork and stayed for a week. Any pleasure I had derived from creating multicolored ribbons of wood with the hand plane was seared off by the scorching heat. One afternoon, the sun and humidity were so merciless that suffering heat stroke in my living room was entirely possible. As the temperature rose, sanding became a harrowing struggle against my own body. My forearms and elbows were painfully sore from the repetitive motion. I wrapped my arms in ice packs and staggered onward. Sometimes as I sanded it felt like they were broken, like I was doing profound and irreversible damage to my arms by sanding all this wood.

The only way I could keep going was to stop sanding and cool off in the bay. I waded in up to my knees at first. The water was crystal clear. Tufts of eelgrass grew up from the rocky bottom. While I stood motionless, thousands of bunker swam past me, their tails brushing my legs in a school of silvery flashes.

Other than a half-hearted and brief foray into swimming lessons at a public pool, I never really learned to swim, and I was incapable of treading water. Growing up on the landlocked plains, I faced more pressing matters on a daily basis, like roping both hind legs of a steer or shooting a pheasant that was flying sixty miles an hour. After almost a week of sanding in the heat, on one of my hourly breaks to cool off I ventured deeper into the bay, and eventually worked up the courage to submerge my whole body. When I dipped my head below the surface, Caper barked from shore. This was our routine every day, Caper and me: a little hotter, a little cooler, a little more sanding.

On many of these sanding days, when I called it quits for the night and crawled into bed, I picked through a book: *The Nature and Art of Workmanship*, a classic 1968 text about craftsmanship by David Pye, a professor of design at the Royal College of Art in London. Pye wrote at length about something he called the workmanship of risk. This was the general act of making something by hand, the creation of a product that could never be the same twice. It was the opposite of what Pye called the workmanship of certainty, wherein objects are made by machines, each one the same every time. Pye used an example to illustrate his point: writing with a pen is workmanship of risk;

modern printing is workmanship of certainty. The outcome of any workmanship of risk is never certain, and its quality is determined by the care, dexterity, and judgment of the maker. Pye argued that "all the works of men which have been most admired since the beginning of history have been made by the workmanship of risk."

The passages sank in deep: my canoe was a true workmanship of risk. I could never build another one like it, even if I tried a million times. The wood would crack in different places, run in different directions, and exhibit different colors. The forms would be pitched at different angles. The whiskey plank and all the strips surrounding it would align and fit to each other differently. My muscles would move in new ways using Dad's tools. Nothing would ever be the same nor could any of it be replicated or perfected.

By contrast, mass-produced products like the orange plastic kayaks that my ex had given me, and which I had purged in disgust before starting my canoe, were churned out by machines—a workmanship of certainty. Even calling up the memory of the kayaks was like dreaming of foreign objects from a different time-space continuum. Beyond my kayaks, though, almost everything we touch in modern American society has been precisely extruded, cut, and assembled by computerized, laser-guided robots. Had I known in advance about workmanship of risk, and the emotional and physical toll of building a boat, I might not have embarked upon the journey, but the hope of one day paddling my own canoe outweighed the risks.

The final week of sanding was a difficult and monotonous toil, interrupted only by the occasional short break to dip in the bay. Ironically, I had gained fifteen pounds but shed two inches off my waist. I felt stronger and more muscular than ever. The cumulative effects of the past year of boatbuilding had given me the kind of strength in my arms and shoulders that I had observed in cowboys, who could effortlessly sling calves onto their shoulders and toss hay bales over corrals. My physical transformation was unexpected. I was more like Dad than I ever thought possible. The person staring back at me in the mirror *was* man enough after all.

Even with tremendous pain shooting into my elbows every time I moved my wrists or fingers, I became capable of endurance sanding. I worked for entire days in a trance, during long pleasant mornings, and lovely swaths of sun-filled afternoons. Sanding had gotten easier for me, but that was different from saying it was getting easy. Wrestling with the fairing board between clenched fists seemed to be one of those tasks that could torture me forever until I had sanded away the whole boat. There came a point during the last day of sanding—the day before Dave's birthday party—when it was just plain hard. My mind shifted into a primal mode as I forced myself to get the job done. I sanded until sanding became unbearable, until I believed that I could not move the fairing board one more swipe, and then I stopped. The hull was as smooth as a frog's belly.

I vacuumed up the dust, stripped off my clothes, and took a shower. Physically demolished, I rested my chin on my chest and stood under the showerhead, watching the slurry of soap, wood dust, sweat, and sandpaper grit flow down my body and into the drain. Then, unable to think about anything else, I did all the tasks necessary to eat dinner as quickly as possible so I could arrive at that delirious moment when I collapsed in bed next to Caper, strapped on my sleep apnea machine, and closed my eyes. That's how I felt by the time the guests started arriving for Dave's birthday party: spent and bored with the canoe, tapped dry of all thoughts and feelings except for the physical relief of being done with sanding.

Dave had hired a caterer to set up a lobster bake on the beach, with layers of seaweed piled on top of rocks in a sizzling and steaming fire pit. A cauldron of boiling salted water awaited potatoes and corn on the cob. Strings of antique filament bulbs draped from the ceiling over my canoe, out the patio doors, and down the bulkhead steps, lined with flickering beeswax candles. Instrumental music emanated from speakers hidden in the hydrangeas. As the food was served and the sun dipped below the horizon, I stood off to the side holding Caper on his leash. He had already been scolded once for stealing a lobster tail off the buffet. I didn't know anyone at the party besides Dave, but I wasn't concerned about that. I had just wanted to help my dear friend celebrate his fiftieth in style. Now, though, all I wanted to do was slink away from the party, lock the door to my bedroom, and sleep for days.

Dave approached me holding a tin camping plate with a steaming red lobster in one hand and a glass of iced tea in the other.

"I've got something to show you," he said, motioning for me to follow him through the sliding glass doors into my living room.

There, Dave's guests had gathered around my canoe. It made me nervous to see them all so close to it, running their hands over it, and whispering among themselves. The only people on earth who had seen my canoe up until this point were Mom, Michael and Glenn, Peter from work, the neighbor John, and of course Dave. Might this large gathering of strangers hold a wild card in their midst, some woodworker who could judge my work as shoddy or inferior? It was a strange experience for me to have so many people crowding my normally reclusive hideaway.

When I walked in the room, Dave's guests peppered me with questions and compliments in rapid-fire succession. Sometimes the music made it hard to hear what they said, but I stood next to my canoe for two hours, absorbing compliments, answering questions, and getting to know Dave's friends in a room full of love, on a night of celebration.

"Your canoe is magnificent! How long did it take you?"

"What kind of wood is this?"

"You have such a talent."

"Is it for sale?"

Then:

"What are those symbols on the chalkboard, is that your logo?"

"Wait, you really built this *yourself*, right here in your house? Is that . . . *normal?*"

I was a bit dazed by the outpouring, and in letting their kind words sink in, my attitude did an about-face. I cared less about my aching muscles and the sanding; I didn't want the party to end.

A huge moon hung low over Peconic Bay, every steel-white pockmark in its face shining clearly as if seen through a telescope. The guests meandered to the beach, where long bands of glowing iridescent blue light appeared underwater in the channel between my house and Robins Island. The extended hot weather through September had not only provided the perfect conditions for ripening grapes, but also for a fall bloom of ctenophores, a nonstinging relative of jellyfish. Each one was the size of a fluorescent blue marble. My canoe-building journey had begun during the coldest February since record-keeping began, when the bay froze over for the first time in forty years, and now here was yet another freak, once-a-century event on Long Island that coincided with the seasonal cycle of my life.

Party guests splashed around in the bioluminescence, some wearing swimsuits, others naked. I stripped down and waded in chest-deep, every movement of my arms and legs highlighted blue. The water was warm even in the night and it soothed my aching muscles. I looked back toward my house, aglow in soft bulbs, where Caper trotted around begging the guests for treats. I felt lucky to be here. What would I do when this was all over, and the canoe was floating, and I didn't have all this woodworking to distract myself?

Dave swam up to me doing the breaststroke with blue ctenophores trailing in his wake. "Thanks again for letting me have my fiftieth here."

"Anything for you—plus, I suppose it was good to push me out of my comfort zone."

"It's also a new life pattern for you," Dave said.

"You mean having people over to my house?"

"No—*accepting love*. You're learning how. The canoe showed you tonight."

"Hmm, I guess so. Hadn't thought of it that way."

I bobbed up and down in the bioluminescence while marveling at the wide sky filled with stars above. The experience felt cleansing. I had seen the beauty of the night stars growing up, riding horseback with Dad on the prairie, but I hadn't appreciated their solemnity like I did now. Seeing the stars made me believe that building this boat was helping me get back to a version of the young man I used to be.

Dave's friends sang "Happy Birthday" to him as someone carried a cake with candles onto the beach. I stayed in the water and joined the serenade, belting out the tune all the way to the final notes: *Happy Birthday . . . tooooo . . . youuuu.*

Right then, while our voices still echoed over the sea, it struck me that this must be what happiness felt like, and that I had finally remembered what it was, and how it could be done.

Later that night after all the guests left, I plugged in my branding iron until it turned red hot, then pressed it against the flat section of hull near the bow. For a minute, the wood steamed and smoked and the pleasant scent of burning juniper filled my house. I pulled the iron off and blew smoke away from the brand. My logo was crisp and black, a permanent scar in the wood, and it looked just right, almost perfect, though perfection was not the point. I relaxed, and tears ran down my face. After the smoke cleared, I wiped the tears away and doused the hot iron with water in the kitchen sink. The canoe stood silhouetted in my house and the filament bulbs overhead flickered, lighting the room just enough to see the object of my happiness.

CHAPTER 28

RASH DECISIONS

The final step in making my canoe watertight was to fiberglass the hull. Who knew that fiberglass, of all things, could inspire joy? I prepped the materials and readied my workspace with music blaring, windows open, my eyes focused on the finish line. Not that I had any clue what fiberglass was, exactly. In researching it, I came to respect it as a technological marvel, but wished I had paid closer attention during organic chemistry lectures in college.

Before fiberglass existed, for hundreds of years wooden boatbuilders relied on fasteners such as copper rivets to hold their boats together. Those boats, however, require constant repair to prevent water from seeping through the cracks in the planking. Wooden boats sealed with fiberglass, on the other hand, do not leak or rot. To make the strongest and most beautiful canoe possible, I would combine the best qualities of both materials: If wood is art, fiberglass is science. If wood is emotion, fiberglass is reason.

My canoe would be draped in a soft and flexible fabric, like a linen tablecloth woven from microscopic strands of glass. Then an epoxy-resin material—a polymer akin to industrial strength liquefied plastic—would be poured onto the glass cloth, soaking through, filling the cracks between strips, and sealing the wood. The finish would be crystal clear, allowing the original woodwork to shine through. After repeating that process on the inside of the hull, I would end up with a laminated wood-fiberglass sandwich every bit as strong as steel, but four times lighter.

I set to work on the first weekend of October, which coincided with Category 4 Hurricane Joaquin making landfall on Long Island. Having started this canoe during a historic blizzard, here I was trying to finish it during a hurricane. Marking time by noting the weather—maybe I had turned into an old farmer after all.

Admittedly, the fiberglass process was intimidating, as I had never worked with chemicals before, but I was ensconced in my house during the storm with all the supplies I needed. I laid everything out on the kitchen countertop. There was the fiberglass fabric itself, plus five-gallon jugs of epoxy and the secondary hardening chemical to be mixed with it. Then there were dozens of mixing cups and sticks, disposable brushes, squeegees, gloves, and cardboard wine boxes flattened to protect the floor. I felt the gravity of the moment in my shaking hands. It was time to make this beast watertight.

I wiped down the hull with a lint-free cloth and acetone to degrease the surface before rolling the fiberglass fabric over the canoe like a shroud. Wearing purple nitrile gloves, I pumped equal parts epoxy and hardener into a plastic cup and stirred it for a minute, then poured it onto the canoe. What happened before my eyes was a revelation. The wood soaked up the epoxy like a dry sponge and its vibrant rust-orange colors popped. Through some miracle of light-bending physics, the once-opaque white fiberglass cloth became invisible. I continued like this, mixing and pouring small pots of epoxy and spreading the viscous, honey-like liquid around with a plastic squeegee. Some of the material dripped off onto the floor before it soaked in, but I kept mixing cups and squeegeeing out the excess until there was a thin layer of epoxy bonding the glass to the wood. It took more than fifty cups and four hours to complete the first coat.

According to the instructions, I had to apply the second coat within four hours of the first, so the two coats would bond to each other. If I waited too long, the first coat would have cured too much to bond with the second, and the glass could delaminate from the wood in direct sunlight. Time was of the essence and I had to keep working. Outside, Hurricane Joaquin stirred up violent four-foot waves that slammed into the bulkhead. Wind and rain pounded against my patio doors and the cloud cover was so dark that it made noon seem like nighttime.

The second coat went a little bit faster than the first, but epoxy splattered everywhere. Thick pools of half-cured goop formed on the boxes on the floor, and every time I stepped around the canoe my shoes stuck to it. When I reached down to peel off the cardboard, I accidentally dumped a cup of epoxy down the leg of Dad's old jeans, which fit me like a glove. Every move I made I felt the sickeningly cold, gloppy, half-cured epoxy touching my skin. I wiped sweat off my forehead; I reached across the canoe; I blew my nose; I went to the bathroom; and every step of the way, inadvertently, I spread epoxy all over my body.

Hours later, I finished the second coat, which seemed like a major victory, except that I was worn-out. I couldn't stop, though. The epoxy was curing fast, and I had to start the third coat right away. The hurricane howled outside over the bay as night fell. I had not eaten a single thing all day. There was no time. By the end of the third coat, I had been working for twelve hours straight without food or rest, but I had to push through one more coat.

The fourth coat became a bewildering fight to maintain my sanity. One pump epoxy, one pump hardener, two pumps epoxy, or was it three pumps? Did I put in four? Or two? I lost count. I threw the cup away and started a fresh one. By midnight I operated with the mental alertness of a zombie.

At some point during the fourth coat, I ran out of nitrile gloves but could not leave the house to buy more, not that the hardware store was open at midnight during a hurricane anyway. I had no choice but to finish the fourth coat without gloves, and by the end of it, my arms seemed to have as much epoxy on them as the boat did.

Sixteen hours after I started fiberglassing, I finished. The canoe was glorious and glowing with the most spectacular russet red and chocolate brown colors I had ever seen. I was too exhausted to care. Caper sneaked in from the den, having taught himself recently how to stand up on his hind legs and push down on the handle with his paws to open the door. He took a few steps into sticky sludge and I shooed him back to safety in the den. When I sat down against the chalkboard wall to admire my work, the epoxy on Dad's jeans squished against the cardboard on the floor. I would clean all that up in a minute. Without quite realizing it or making a conscious decision, I closed my eyes for a harmless little snooze.

Six hours later, I woke up stuck to the floor.

While I had slept, the epoxy cured and bonded together everything that it touched. My jeans and shoes stuck to the floor so tightly that I had to unzip and undress myself before I could stand up. In doing so, my jeans remained stuck to the floor and I ripped some hair off my legs and groin—a partial Brazilian wax. The hair on my arms stuck to my skin and my left eyelid was fused shut. One of my shoes had been epoxied to my sock in an immovable plastic clump, which I pried off with tin snips from Dad's toolbox.

The creeping realization began to sink in that I had done something horribly irresponsible by allowing the epoxy to cure on surfaces other than the canoe. Hobbling around with one eye shut, I ran back to my bedroom and found Caper sleeping peacefully on my bed. When he jumped down to greet me, the bedsheets came along with him, epoxied to his fur. I cut him free and threw the sheets and all the cardboard boxes out into the hurricane-soaked yard. I got in the shower with a jug of white vinegar and aggressively washed my skin, hoping that the vinegar would neutralize and break up the epoxy. I scrubbed until my skin was red and inflamed, but the epoxy had already cured into an inert solid.

After the shower, I became dizzy and a faint pinprick sensation radiated down my arms. At first it felt like innocent mosquito bites, but over the course of the next few hours, every inch of my skin ignited in a searing, red, crusty, oozing rash that I scratched until it bled. My breathing became constricted and wheezy as I stumbled into the kitchen and flipped frantically through the pages of *Building a Strip Canoe*. One disturbing sentence from the safety chapter stood out from the rest: "Even minor skin contact with epoxy can cause sensitization and chronic health problems, including chemical burns and allergic dermatitis."

I threw down the book and drove myself to the urgent care clinic, where I stood naked in front of a doctor, explaining my predicament.

"Did you roll in poison ivy?" she asked the second she laid eyes on me.

"I wish. I was fiberglassing a boat."

"Hmm, that's a new one," she said while snapping on a pair of gloves. After prodding and poking she informed me that I had a severe case of allergic dermatitis and some chemical burns. I might also have been at risk for infections since I had scratched myself raw. The doctor administered two deep-muscle shots, Prednisone and Benadryl, to slow my body's histamine response.

"How long before this clears up?" I asked.

"If you don't reexpose yourself, a week or ten days," she explained. "But this means your body is sensitized to whatever chemicals you were using. So, if you touch that stuff again, your reaction will be worse every time you're exposed."

"But I can't stop boatbuilding, I mean—"

"Is it your job? Does your livelihood depend on it?"

"No, I have a day job, this is . . . I don't know, I guess it's a hobby?"

"Then I'd say you should find a different hobby, at least one where you're not handling fiberglass."

"I can't do that. I have to finish this canoe," I said, my panicked voice rising in pitch.

"What's more important, the canoe or your health? Don't touch fiberglass again, or I'll see you back here for more shots."

"You don't understand, it's not *just* a canoe, it's . . . you see, my dad—"

"Look, if you'd burned your hand holding it over a candle, I'd tell you to remove your hand from the flame. Pretty simple."

"So you're telling me I'm allergic to my canoe?" I wondered aloud, afire with humiliation and fear. It was the kind of question I could ask but did not want to have answered.

"In a way, yes," she said.

The doctor left me alone in the room to get dressed and consider what had just happened. Despite all the things I had endured trying to build this canoe, I had never considered quitting for real—like actually giving up and calling the whole thing off for good. I shuddered at the possibility of calling Mom and saying it was an ambitious idea that was not meant to happen. I slumped down on that crinkly white tissue paper lining the padded doctor's table and feared the worst: if I couldn't finish, I would fall short of Dad's legacy. I didn't see a way around it, but I was desperate to find one. I felt an odd mix of both terror and relief: at least if I did quit, I would have a good excuse.

I went back to my house and lay in bed, zoned out on the painkiller and antihistamine. After I changed the bandages on my chemical burns, I wrapped myself in so many layers of antibacterial gauze and blankets that I looked like a cocoon with feet. That day and the next three or four days, I flopped around in bed wearing soft pajamas, convulsed by fits where I was so stung by the rash that I would happily have thrown myself off a bridge to make the itching stop.

I began having deranged fever dreams. I repeatedly found myself back in South Dakota, running away from charging bulls and rattlesnakes. But mostly I hallucinated about Dad. In my fog of aches, groggy and shivering, I had a startling idea: *I'll call Dad.* It came to mind as if his death were yet to happen and there was time to stop it if I could get him to answer the phone. *I must call Dad and tell him about my canoe.* In my haze I would be standing in the doorway of Dad's shop, the windowsill piled high with white sleet, the floor littered with hoof trimmings, the whole place smelling like his cologne, still warm and alive with his presence. Then a sudden sharp flash flung me back to real life, to ordinary time, and I became aware of the lights in the living room and the waves crashing on the beach. Caper, velvety and snow-white, hopped up on the foot of the bed and nestled down for a nap between my feet. Before long, he was snoring blissfully. I smiled foolishly at the imagined smell of Dad's shop but then caught myself and snapped out of it. He would never answer the phone.

I worried that my fiberglass epoxy disaster had damaged more than my skin. Maybe my brain and nervous system had been damaged, too. And if I couldn't finish the canoe, maybe even my soul would be damaged in some irreparable way that might not be readily apparent until many years from now.

The very thought of abandoning my project filled me with shame and frustration. Until my rash healed and I came up with a solution, I couldn't bear to look at the canoe. I searched the internet for a storage unit in town and found an old barn a couple of miles from my house, next door to the same café where I took Mom for dinner the night before Caper ran away. The barn was built in the early 1800s and used to be the village blacksmith shop, but it had been sitting vacant for some time. When I felt well enough to get dressed in normal street clothes and leave the house, I signed a lease and asked Peter from the winery and John from next door to help me move the canoe to the barn.

On the count of three, we deadlifted the strongback. The Douglas fir beams dislodged from the floor: crackling, groaning, popping, and tilting sideways while we fought to gain control. It sounded like we had uprooted a tree. We held it there for a moment, suspended at waist height, shellshocked by its beauty and magnitude. Even for three grown men, it was a lot to handle. The colors and light reflected off the surface and illuminated Peter's and John's faces. We slowly maneuvered the canoe past the taxidermied duck, out the sliding glass doors and onto the brick patio. I was not wearing gloves, and the strongback gouged into my hands.

"Wait, I need to adjust my grip," I said. We set the canoe down on the grass near the bulkhead and it seemed to come alive, framed by the blue water in the background. That was supposed to be its destiny. It was the oddest thing to contemplate this beast not being here anymore. I could tell the guys to turn around and take it back inside, but then what? Sharing my living space with it only reminded me of my failures. I pressed on, practically panting with disappointment.

We hoisted it back up and walked slowly around the side of the house, toward the vineyard truck in the driveway. Time slowed down. The hull quivered slightly with each excruciating step, bending but not breaking. We heaved it into the back, sliding it up toward the truck cab and securing it with ropes. I could not help but feel like I was a pallbearer at Dad's funeral, sliding his casket into the hearse.

We unloaded the canoe into the storage barn and locked the door. I stood for a long while not ready to leave it there. I didn't want to, but I needed the separation.

"Why did you do that?" Peter asked, somewhat alarmed as we drove home.

"I'm allergic to fiberglass. Don't really know if I'm going to finish. Need a break."

"Then pay someone else to fiberglass it!" Peter said.

"No, I have to do it all myself. That's the promise I made in the beginning."

"Well, I know I teased you, but I didn't *really* mean you should take up macramé," Peter said. "I was kind of rooting for you, hoping you'd finish."

I stared out the window at the gray October sky as we passed pumpkin farms full of day-trippers from the city, who strapped corn husks to the roofs of their Range Rovers. I was so dejected that I didn't say another word to Peter.

He dropped me off at home and I went for a dip in the bay. The salt water stung my rash and eyes, but it was the only thing that kept my emotions from completely unraveling. I floated facedown and opened my eyes underwater. The eelgrass wafted in the current. Something shiny and silver appeared in my peripheral vision. Of course, Dad's hammer was long gone by now. This was only a school of bunker. A hermit crab dragged its shell across the bottom as if in slow motion, like an astronaut on the moon, both heavy-shelled and weightless creatures, finding their own slow way, stirring up silt as they moved. My frayed nerves were calmed by the gentleness of time underwater.

I resurfaced. The cool southwest wind blew toward me from the direction of the scrub brush on the sand dunes, wafting the scent of rose hips that had split open and fermented in the October sun. Maybe I had it all wrong. No matter what I had set out to do in my life, in the scale of the universe, my problems were no more significant than the flick of a bunker's tail or a rotting rose hip.

The dock light on Robins Island twinkled, taunting me. I was mad at myself for having come up with this outlandishly stupid idea of building a boat. Elsewhere, my friends had had their easy summers, sipping cocktails on the beach and enjoying time off from work while I had toiled on my canoe every waking minute. It was all a waste of time. Medical reasons or not, to hear myself mulling over the words *I quit* made my throat constrict with sadness.

I trudged up the bulkhead into the house. In the living room was a canoe-shaped stain outlined on the floor, consisting of dried glue, sawdust, and epoxy—the ghost of a boat that almost came to be. It was such a strange sight. I moved Dad's toolbox from the workbench to the floor, centering it in the phantom canoe. Unable to resist, I turned, reluctantly, to his tools. They fit back in his toolbox easily. I couldn't bring myself to clean them. My heart seemed to cleave in half each time I grabbed a tool and dropped it in. The last thing I tucked on top was Dad's rodeo rope, which I had to stuff inside and then sit down on the lid to contain. With a twist and a click of the bronze latch, the toolbox was closed for business.

I picked the few straggly cherry tomatoes from the patio garden that had been nourished all summer by sawdust. I ate them slowly and savored them. The sun was almost down, and with November right around the corner, the days would only get shorter and colder and the plants would wilt. Soon it would be winter, and the anniversary of Dad's death would come and go without a canoe to commemorate his passing.

While still chewing the tomatoes, I stood in the ghost canoe outline and faced the taxidermied duck. I said Dad's name out loud. My mouth shaped the word.

"Leon," I said. It sounded so weird coming from my lips, as if I were having a ceremony for him. The beady black eyes of the taxidermied duck stared back, lifeless.

Saying *Leon* instead of *Dad* felt like the first time I acknowledged he was more than just my father. When he died, I lost that man, too—Leon, the man who was like me with his aches and pains, imperfections, triumphs, and disappointments, and dreams that never came true.

He was so proud of his name that he had given it to me as my middle name: Trent Leon Preszler. I had stopped using my middle initial in my twenties. It represented who he had not been to me and I wanted no part of it. My whole adult life, that black hole between my first and last names was the thing that had always been missing. It was the void I had tried to fill with material possessions and professional accomplishments, but nothing ever made the emptiness go away.

I tried looking at the wood duck without being embarrassed that I couldn't finish the canoe. I slumped on the floor, buried my face in my hands, and began to cry, or half cry, with ugly dry choking. Caper pawed at my shoulder and snuffled anxiously against the back of my neck. It seemed I had spent the whole year getting this canoe started, and now suddenly, it was over.

CHAPTER 29

THE HEARTS OF MEN

Autumn came quickly. The trees changed color and frost accumulated on the ground. Harvest at the winery churned along. I woke up one Tuesday in late October and rubbed sleep from my eyes. I didn't drink the night before, yet for some reason, I felt miserable and hungover. Of course I did. It was the fifteenth anniversary of Lucy's death.

WHILE I INHERITED Dad's hands and stubbornness, among other things, Lucy had inherited a terminal brain disease. The doctors finally diagnosed her after years of mysterious ailments and meetings with specialists. We were told she had a child-onset degeneration of neurons in her brain that affects roughly three out of every million people—something so rare that the medical community hadn't coalesced around a proper name for it, other than its technical definition. A poor rancher's daughter from flyover country didn't stand a chance at raising awareness, either. To my knowledge, there had never been a black-tie gala in Rockefeller Center to raise money for a cure to olivopontocerebellar atrophy with spinocerebellar ataxia type 3.

Healthy males with this disorder mapped onto their X chromosome pass the mutated gene to their daughters, who have a fifty-fifty chance of being a neutral carrier or falling victim to it. Because the diseased gene was X-linked, Dad could not have passed it down to me; males only pass their Y chromosome to male offspring. Odds were good that the gene mutation happened when he was exposed to Agent Orange in Vietnam. I would never know if he blamed the U.S. Army for Lucy's disease, or if he even characterized it like I did, as her inheritance.

Before I moved to New York for grad school, I visited Lucy at Children's Care Hospital and School in Sioux Falls, the only place in South Dakota where someone with her condition could be treated. When I walked into her room, she was in her adjustable bed trimmed by a runner of quilted storage pockets and surrounded by steel grab-bars wrapped in foam padding. She saw me and let out a groaning sound.

"What is it, Lucy?" I asked, walking to her bedside, but as usual, I could make out nothing of what she was trying to say. "Was there something?"

Her twisted mouth moved into what I had learned to interpret as a smile.

"Oh yes, I see!" I said, always responding like I understood her. I had visited often over the years, and each time my playacting became more convincing.

I turned down the top of the pink quilt Mom had sewn, damp with drool and sweat, and scooped up Lucy's wasted body. I drew her toward one side of the bed to make room for myself. Her eyes, buried alive inside her body, followed me as I lay down next to her.

"Tell me about everything. I heard you have a pet hamster that eats cardboard," I said, but of course she couldn't tell me. She made that gurgle in the back of her throat that could mean anything.

I reached for her hand and we held still. Her palms were tiny, and her fingers contorted around themselves. Muscle wastage had left her forearms looking gaunt and sinewy, like twigs of bleached driftwood. Her elbows flared out in swollen knots. Her shoulders convulsed involuntarily. We lay there for a long time, staring at the construction paper cut-out hearts and rainbows taped to the ceiling. Her legs started to spasm. She needed to use the bathroom. I called in the nurse, who pressed a button that brought the bed upright to a sitting position. It was time for me to go.

"Hug?" I said. Lucy made a high-pitched sound that I equated to giggling.

I leaned in as far as I could, bending over so the nurse could place Lucy's shriveled arms around my neck. I felt the bedsores on her protruding shoulder blades. After a minute, I wiped drool off her chin with my shirtsleeve and stepped back. The nurse detached Lucy's stomach catheter feeding bag from the side of the bed and prepared to lift her into a wheelchair with the assistance of a complex array of straps, winches, and pulleys. There was always a moment, as she floated from bed to wheelchair, when the whole precarious system wobbled, and I got nervous that she would drop down and smash her tailbone. Today's ride was smooth, though, and Lucy stuck the landing. The whole ordeal took fifteen minutes. The nurse rolled the chair onto the tile ramp by the bathroom's double-wide doors and turned around to face me.

"Okay, Lucinda, time to say bye-bye to your brother," the nurse said while binding Lucy's body with a seat belt.

Her eyes howled up at me from the mechanical chair. Her warped mouth, stiffened in terror, looked like the expression on a Greek tragedy mask. She rocked her wilted torso back and forth before squeezing out a familiar syllable from her rigid lips.

"Tent!" without the *R*. I smiled to put her at ease but hearing her say my name chilled me.

I longed to know anything that might be going on inside her mind, if she was in pain, and how I could make her comfortable. My wanting to help didn't matter, though. The result would be the same. The cold truth I had come to accept was that I could never know her personal experience, so I projected my own to fill the gaps, in hopes that she might see a sliver of light.

"I know, Lucy, but I have to go to grad school in New York. I'll come back to visit. Take good care of that hamster, okay?"

Her mouth twitched a little. Was she trying to say it was exciting that I was moving to New York?

Her eyes moved toward me. Did she just say *I love you* in her mind?

Her movable hand lifted an inch and dropped in her lap. Was she trying to wave goodbye?

I didn't know, but I wanted it all to be true.

"Yes, it's exciting! That's so sweet of you, Lucy. I love you, too."

She strained her chin up and looked at me. More than her rigid limbs or shattered speech, her eyes hurt me deeply. They were brown and clear and full of life, even when her body was not. When I looked into them, I thought her eyes were saying to me that nothing on this earth was bearable anymore, but that reaction was as much about me as it was about her. By bearing witness to her suffering, I couldn't help seeing the world from her perspective.

The nurse wheeled Lucy into the bathroom. I had no idea that would be the last time I saw her alive.

I CONVINCED MYSELF that there was nothing I could do about Lucy's condition and I had to get out there in the world and live my life apart from her. A whole new chapter started for me when I got to grad school, and I poured myself into it. The first time I saw Cornell University, I knew I wasn't dreaming because I had never dreamed of anything so majestic. The clock tower made of hand-hewn stone, and beyond that, perfectly clipped grass amid brick buildings with Greek columns the color of beach sand—all of it was arranged in quadrangles surrounded by emerald-leafed trees, perched on a limestone cliff overlooking Cayuga Lake. I ate breakfast and lunch in a great hall with ceilings so cavernous and stained-glass windows so colorful that it felt like eating in a Gothic church. Beautiful people walked to class and laughed, with luxurious brand-name backpacks slung over their shoulders. In my free time, I explored the most extraordinary collection of campus trees in America, photographing the thirty-nine varieties of pine, twenty-five types of spruce, fifty different shapes and sizes of oaks, and twenty cultivars of walnut. Cornell was the antidote to the treeless prairie.

There were days in my first semester when my luck scared me, and I felt unworthy of the College of Agriculture. When other students and professors asked where I was from, I said, "I'm from South Dakota," a phrase that always elicited the same response: "I've never met anyone from there!" I enrolled in classes that I had craved for years, grateful for the opportunity to learn. Cornell paid my tuition and lodging and in exchange I taught undergraduate seminars in botany and agricultural business, things I would gladly have paid *them* for the privilege of doing.

My personal life took a different turn as well. I fell in love with straight men all the time, chasing after guys I knew were off-limits—guys who were roughneck, unforthcoming, and silent. I would see a midfielder from the Cornell lacrosse team in the library, and before I had spoken one word to him, I had already convinced myself that we were fated to spend a wonderful future together. I learned that straight guys would do things they claimed they wouldn't do in a million years if we went bar hopping in Collegetown after class. Sometimes they would regret it in the morning and disappear quietly from my bed. If they stuck around for breakfast and wanted to talk about farming, I might see them again.

After one particularly upsetting breakup, I went to Cornell's student health center and talked to a counselor, a gay man in his sixties with compassionate eyes and a soft, gray beard. He told me that since my father never said the words *I love you*—and since Lucy couldn't physically speak words, yet expressed her love for me in other ways—as an adult I was conditioned to experience love as a silent expression. The counselor said I subconsciously sought out taciturn, unavailable men so I could receive the sort of fatherly love that I missed out on growing up. I stormed out of his office, denying that anything he said was true. I was still a stubborn kid from Faith underneath the Ivy League facade.

One of those closeted men stuck around long enough to become my first steady boyfriend. His name was Tim, a tall Texan who spoke with a drawl and wore polished boots, from the Texas A&M Corps of Cadets. We had met the previous summer in Washington, D.C., when I was an intern for President Bill Clinton at the White House Office of Science and Technology Policy, and Tim worked on the presidential campaign of George W. Bush. In our private moments, we joked that we were a bipartisan power couple, like the gay version of Mary Matalin and James Carville. Tim and I were gearing up for an autumn weekend of homecoming football and apple picking when Dad called and left a nine-word-long voicemail.

Hello, Trent? It's Dad. Lucinda died. Better come home.

Tim and I flew to South Dakota together. I told my parents that I was bringing my roommate for emotional support. We arrived at the church and I casually asked Dad if Tim could sit next to me in the front row.

"That's reserved for family," he said.

"Okay, no problem," I said with an indifferent shrug, swallowing any scrap of emotion I might have felt about his answer. Tim sat in the back row.

First there was a breath of the organ, then a whisper from the minister, and the whole congregation joined in a chorus of mismatched voices raised in mournful worship. During the singing of the Common Doxology, the most familiar hymn of my childhood, the pallbearers wheeled an open casket containing Lucy's body up the aisle. I looked at Tim in the back row and wanted desperately to be holding his hand or crying on his shoulder, but I had to settle for the distant, silent kind of love.

Mom reached her delicate hand over to touch Dad's calloused hands. The gesture startled me; they never showed affection in public. Dad covered his face. Choking sobs came from the other end of the front pew. I was twenty-three, and it was the first time in my life that I had seen Dad cry.

The minister directed us to fold our hands in prayer. While everyone else bowed their heads and closed their eyes, I stared at Lucy's corpse on display before us. With her face relaxed for once instead of being tortured by brain and muscle spasms, she looked peaceful—less like my disabled sister and more like a beautiful young woman. I wept for the best friend I had lost.

By the time we reached the cemetery, Tim had grown so uncomfortable with the whole situation that he stayed in the car while I attended the burial rites. It was probably for the best anyway. Had Tim been next to me at the gravesite, I might have slipped up and accidentally held his hand or shown affection in a way that would have triggered alarm among the congregants.

I couldn't say goodbye to Lucy. I didn't want to leave my twenty-five-year-old sister trapped in the cold October ground. Dad had to pry me off the casket and drag me to the car. After the burial, Tim and I had a brief and wordless interaction in the parking lot of the cemetery. I got in the car with my parents, and he left for the airport. We didn't even get to say a proper goodbye.

Everyone gathered in the church basement for fellowship, which, in Lutheran custom, was the occasion when women with German last names served paper plates of red Jell-O, sliced ham, and potato salad. Fluorescent strips buzzed overhead, casting shadowless cold light on the cinder-block walls. Sitting across from Dad in a plastic folding chair, I wore dark aviator sunglasses and picked at a piece of ham with my fork. I had no appetite and was appalled that anyone expected us to make small talk.

While Mom was in the church kitchen with the other women, I worked up the nerve to tell Dad the most terrifying secret of my life. Neither Tim nor I were out of the closet. We had told some friends, but not all. My feeble attempt at coming out was part of the messy phase of believing in who I was on the inside while at the same time trying hard to project confidence in a specific identity to the outside world. I was a few years behind the curve, having lacked any of the opportunities straight kids had to iron out these awkward details as teenagers. Growing up, I was jealous of the guys who escorted their girlfriends onto the field at halftime of the homecoming game while an entire football stadium validated and applauded their sexuality. As for Dad, I was terrified that he might punch me, or accuse me of trying to steal the spotlight from Lucy, or tell me that I would burn in hell. I was filled with a combination of hope, anxiety, and a foreboding sense of possible loss if Dad rejected me—all set against a backdrop of devastating grief for my sister. On the other hand, I figured if there was any moment where he might look on me with kindness it would be now, when all we had was each other. My hands started shaking, and my voice cracked as I stammered through the words.

"Dad, um, Tim isn't my roommate, he's my . . . friend, but a different kind of friend, like, he's a guy, right, except he's also my friend that I sometimes spend the night with, in the same bed, know what I mean? So, I just wish he could've sat next to me at the funeral, okay?"

Dad's facial expression was blank and steely, and his response was cold: "Let's not talk about *that* again."

I clenched my jaw. I wasn't going to make a scene right there in the church basement. Lutherans would never be so dramatic as to confront a situation head-on or cause a ruckus that upset people. We preferred a passive-aggressive slow burn. I honored Dad's request and kept my mouth shut.

A few weeks later, just as I was getting back on my feet at Cornell, an envelope arrived from the Church of the Lutheran Confession—the home church where we buried Lucy, and the most conservative orthodox sect of Lutheranism. I tore open the letter, feeling numb, as if I were about to read my verdict after the judge had already ruled on my case. In it, the minister described the church's adherence to closed communion, which limits all forms of religious fellowship to Christians who express full agreement with Scripture. And full agreement, he explained, requires rejecting as sacrilegious any effort by which the intellect of man would disparage God's literal inerrant word. That part of the letter seemed a little harsh, but not too terrible, all things considered. I continued reading and discovered that my name had been removed from the church membership. In classic passive-aggressive Lutheran style, there was no explicit reason cited as to why or how this decision had been made, or by whom, but I could only assume it's because I was gay.

Sins of the flesh were rooted out with vicious efficiency in our church. One time when I was in high school, the minister forced a woman to stand up in front of the entire congregation on Sunday to confess that she was pregnant out of wedlock—a real modern-day Hester Prynne. The sound of her sobbing while clutching her pregnant belly was seared into my consciousness and served as a stark warning about the dangers of lust. The church's stance on homosexuality was certainly no secret, but I had to find it for myself so I could absorb the words—not as abstract ideas, but for the first time as rules that applied specifically to me. My suspicions were confirmed when I found on the internet the Church of the Lutheran Confession's official position statement on homosexuality.

> *God did not create people as or to be homosexuals any more than He created people as or to be thieves, child abusers, or alcoholics. . . . Those who claim that God made them as or to be homosexuals are committing blasphemy against the holy name of God. . . . People who have homosexual tendencies may be members of our church so long as they try to overcome their condition, and are not practicing the sins associated with such a lifestyle. . . . Concerning our position on homosexuals, we choose to face the wrath of the world in time, rather than the judgment of God in eternity.*

The letter also provided the name and number of a Lutheran minister in Buffalo, New York, in case I wished to redeem myself through private Bible study—with a goal, I assumed, of becoming a true, God-fearing heterosexual.

Excommunication was the knockout body blow, the ultimate rejection. Everything was ripped away on a whim. Suddenly, any sense of confidence that I ever had evaporated. My whole life felt conditional. There was no worth inherent in who I was as a person, neither from Dad's nor the church's perspective. Any worth had to be bestowed upon me through rules that were written specifically to exclude me—the ultimate catch-22 situation. While the church believed that I was an abomination, I knew it was not that I had done something wrong, but rather that I existed in a fundamentally wrong way. There was something flawed and incurable about me. I wasn't the man they wanted me to be, even though they always said from the other corner of their mouths that God loved everyone and created us all in His image.

I had grown up with the story of Jesus Christ as a shepherd watching over His flock by night, loving all sheep equally. When you were on the outside looking in, though, and only part of your whole person was accepted, everything was different. Those comforting parables about Jesus's unconditional love didn't apply to gay people. I was wrong to believe that the church was a place of kindness.

There were words I had never expected to read, and when I read them, they froze me, and I couldn't move for a while. I was so paralyzed that I almost flunked out of Cornell. In fact, I did fail agricultural economics. I was up all night crying before the final exam and didn't even answer half the questions. I deserved an F, but the professor felt sorry for me and gave me a C.

"You were dealt a bad hand in life, it's the least I could do," she said.

Whether Dad loved me or not, he didn't love who I turned out to be. In Christian parlance, that essentially boiled down to the saying *love the sinner, hate the sin*. I didn't want the "hate the sin" part, the caveat, the asterisk, the footnote. I didn't want to know or hear that this was what responsible adults thought about their own children, that they valued a religious sect's interpretation of a two-thousand-year-old book more than their own flesh and blood standing right in front of them. I was sad that this separation would never quite be bridged but it was a circumstance we couldn't avoid. The legacy of our place in the world was that Dad had been brainwashed by the Church of the Lutheran Confession to believe that his only begotten son was an enemy of God.

Freed from the long leash of Dad and the church, I saw how I didn't belong and could never be forced to belong. Being gay was just who I was, and I didn't need to keep my true self hidden. Once I was able to embrace that truth, I saw how sad it was that I had surrendered so much power to Dad and the church. The greatest insult I could ever bear was that I had allowed them to define me to myself.

Dad never spoke about my sexuality again—not even once, and not even to Mom. It would be another three years before I told her I was gay, too. At some point she asked me about my old friend, the tall Texan named Tim.

"Did he ever get married?" she said.

"No. He's gay, Mom, and surely you must understand I am too, after all this time."

"I just hope you know what you're doing," she said, before telling me that Dad could never accept it, he believed it was a sin, and he would not speak of it with me again—not now, not ever.

"You guys must have talked about it at home though, right?" I asked. "Just the two of you?"

She didn't answer.

It was a truth I wished I didn't know. They couldn't talk to me about being gay or speak to each other about who I was because the church condemned people like me. I was an issue to be avoided, a problem to be swept under the rug, a voice to be silenced until I became invisible.

Over time, my understanding of those events shifted as I swung into adulthood and my career blossomed. I came to see how all my decisions added up to make a life, not unlike trees adding a new growth ring every year. The lessons and secrets from early days were stacked inside the wood, many rings deep. I had often wondered how my life would have been different if I hadn't dragged a closeted boyfriend home for Lucy's funeral. Depending on the day, I swung from wanting to apologize to the absolute conviction that it was the right thing to do.

Life got better after that. I was finally able to be who I was and to have less fear about myself. I answered the question of whether I deserved to exist by coming out of the shadows and working harder. I pegged my self-worth to my own achievements, not to Dad's approval or the church's. I took pride in being an out, gay, working professional, and my confidence grew. There were a handful of occasions in my twenties when I put on a happy face and attended family gatherings and reunions around the country. I assumed that because I had changed, Dad would have, too—and maybe things would be easier with aunts, uncles, and cousins around as buffers. But Dad and I were like oil and water, and we took great pains to avoid eye contact and conversation. Nothing would ever be the same between us. Rejection was a two-edged sword. There was a fork in the road, and I followed the path that led away from Dad. I holed up in New York and drifted from my family, year by year, little and often.

I never wanted to hurt my parents or make them feel uncomfortable for their religious views. That had always been my excuse for not making more of an effort to force them to really know me, to force them to share a little more of the substance of our lives. When the topic of my sexuality, of who I really was, finally came up in a meaningful conversation after so much silence, I was thirty-seven years old and Dad had a week to live.

AFTER THE ANNIVERSARY of Lucy's death, autumn rolled on. The days were cool and empty, filled only with the sounds and smells of the vintage. Night after night, I came home from the winery and dreamed about paddling my own canoe. I seized on the image of myself on the water, took it in, and comprehended it all at once, and found that it came accompanied by a fierce determination, a sense of rising resolution. I was fed up. I had had enough of the pain. I was tired of finding myself in this position of suffering, being scared and hurt, and endlessly asking myself what was wrong with me. I could no longer depend on anyone else to find my sense of who I was in this world. The canoe contained every scrap of love that I had ever lost or found. The decision was mine whether I loved myself enough to finish it.

CHAPTER 30

ASHES TO ASHES

One time after we sold our cattle at the Faith Livestock Auction, Dad handed me a crisp five-dollar bill and told me to buy anything I wanted at Faith Lumber Company. After perusing the racks of candy and petting the baby rabbits and chicks, I picked out a pair of extra-small leather gloves to match Dad's. He ruffled my hair and told me I made a good choice, then added this: "Your hands are your most valuable tools."

It seemed so logical—of course my hands are important—but in the case of fiberglassing my canoe, I had been too brazen in my ignorance of safety precautions. I didn't respect my most valuable tools, and I took my good health for granted. Being allergic to epoxy didn't have to mean that finishing my canoe was out of the question. I simply had to take extraordinary steps to protect myself when I was ready to fiberglass the inside of the hull. So that is how I found myself wandering the aisles of the local medical supply store searching for the right gear.

Peter and John helped me move the canoe back from the storage barn into its rightful place in my living room. I loosened the station forms from the strongback, and they helped me flip over the hull, so the canoe was right-side up. The inside was stained with glue and epoxy that had squeezed through the cracks between the joints. It was nothing that a little rasping, planing, and sanding couldn't clean right up. I spent the next two weeks fairing the inside of the hull, and when I was ready to fiberglass it, I dressed up in a full-body hazardous materials suit, which included a helmet, sealed face guard, one-piece body suit, and gas mask with a respirator. I mixed and poured cups of epoxy to soak the fiberglass cloth just as I had done on the outside of the hull, except this time I resembled a character from a science fiction movie leaning over my canoe as though it were an alien life-form. When I finished the four coats, I stripped off the hazmat suit and rejoiced in victory: not a single drop of epoxy had touched my skin.

It was now early November and I had just over a month to finish. While the epoxy cured for a week, I brushed up on the remaining steps of the build. There were several parts and pieces to be attached to the hull that would contain and control its springlike inner tension. Long strips of hardwood attached to the sheerline would form the top railings, called gunwales, pronounced like the word *tunnels*. They stiffened the hull and held it together if a strip were to break upon hitting something in the water. Thwarts were crossbeams used to bind the gunwales together at a fixed width spanning the opening of the canoe. The gunwales met in a point at the bow and at the stern and were connected by triangular decks. So, the gunwales, thwarts, and decks provided built-in redundancies that gave canoes their strength. They held all the dynamic forces of the wooden spring in place.

I had been so focused on the strip-planking and fiberglass details that I hadn't even considered the time and skills required to build these other pieces. A quick search of the internet showed that I could easily purchase the thwarts, the decks, and even the seats premade. Of greater concern, though, was finding white ash lumber that was long enough to make the all-important gunwales.

Ash is a hard and sinewy wood typically used for baseball bats, and it was the preferred wood for boat gunwales, too. As a living tree, ash was known to be a tough and resilient species, one of the most cherished American hardwoods. I had always loved its compound, fern-like leaves, how they feathered the light and made life feel softer than it was. Its tapered seeds even resembled miniature sailboats.

I called Roberts Plywood and three other lumberyards in the New York metro area, but none of them carried ash in the right dimensions. It was one of the most common trees in America, but there was a severe lumber shortage related to an outbreak of the emerald ash borer, a beetle that slipped into the country from China on a pallet delivery in 2002. Within a year, the parasite had already killed dozens of trees in Michigan and released hordes of larvae that found perfect living conditions in the moist green layer just underneath the ash tree bark. City gardeners mobilized a counterattack, lopped off infected trees and burned them, and doused the living ones with pesticides, but their efforts were merely a drop in the bucket. Death raced across Michigan and Ohio, jumping dozens of miles a year. Trees succumbed by the hundreds of thousands. Botanists across the nation watched dumbstruck as America's noble ash population crumbled. A decade after the beetle first appeared, tens of millions of trees were dead. Without a practical way to kill the beetles on a large scale, scientists have estimated that all eight billion ash trees in North America will perish by the year 2030.

I made a round of frantic phone calls to lumber mills upstate and couldn't find one that carried long-grain boatbuilding ash. There were still a handful of small wooden boatbuilders in the Adirondacks. If I could track down one of them maybe I could find out where they sourced their lumber. Repeated attempts at contacting them were unsuccessful, though. Time was ticking and I couldn't wait on someone else's schedule to start making my gunwales, so I loaded Caper into the car, and we drove north to the Adirondacks in search of wood.

The farther upstate I drove, the more I saw evidence of the emerald ash borer's ravaging of forests. Miles of hillsides had vacant pockmarks where once grand ash trees withered and died. There was such an abundance of cheap firewood that mountains of it were stacked alongside roads and farmhouses. With so many diseased forests and not enough logging companies to process it all, trees had been left to rot where they fell. Their bark had peeled off in grotesque sheets revealing snakelike channels where the insects had burrowed. I was, outraged by the injustice that these losses could never, ever be reclaimed. I was witnessing the extinction of a species, but not just any species: ash was the wood Dad had used to build our toboggan.

Boatbuilders jealously guarded their sources of prized lumber, but I visited two boat shops along Lake Champlain, explaining my desperate hunt for ash. They both graciously volunteered the same advice: go see Preston at Essex Industries. I had never heard of the place and it didn't come up on any of my internet searches, but that kind of under-the-radar discovery was exactly why I drove all the way up here.

I pulled up to a secluded parking lot in front of a repurposed schoolhouse and the thought crossed my mind that this could either be a lumber mill or the hideout for an armed antigovernment militia. Inside, I was greeted by a young woman sitting in a wheelchair at the front desk.

"Um, I'm not sure I'm in the right place. I was sent here for boat lumber," I said. "Is Preston around?"

I watched the woman release the brake on her wheelchair. Her movements were jarring and uneven. She seemed to think for a second about adjusting the lever, then furrowed her brow and did the movement she had thought about. After she was done making this concerted effort to accomplish a basic task, the expression on her face lightened and she looked up at me. I couldn't help but think of being a kid and watching helplessly while Lucy struggled to push her wheelchair.

"You're in the right place," she said with an enthusiasm that caught me off guard. "I'll find Preston."

While I waited in the cinder-block hallway a young man with Down syndrome walked up and introduced himself. He adjusted the strap on his safety goggles, and he was also wearing a shop apron with a measuring tape and pencils tucked into various pockets. I began to wonder what the odds were that I would see two people with disabilities in one day, in the same building, in the middle of the Adirondacks.

Preston, the manager, came through the doors and greeted me. He was a burly, heavyset man with a booming voice and military-style flattop haircut. His khakis were pulled up just a bit too high, the same way my high school gym teacher wore his pants, and I half expected to see a whistle dangling from his neck.

"I'm told you're the guy to talk to about boatbuilding ash," I said.

"Ah yes, you're in luck, I've got a stockpile of beetle-free lumber in lengths up to thirty feet," Preston said. "It's as tough a wood as you'll ever find around here, which makes the beetle kill ironic."

I nodded. "It's like cancer, the beetle doesn't care how tough the tree is."

He walked me through the processing area wearing a hard hat, safety goggles, and earplugs. One wall was stacked floor to ceiling with lumber: ash, black cherry, and white cedar, all of it harvested locally. Preston explained that this was a work-to-live facility. He hired people with mental and physical disabilities, recovering addicts, and people who just had had a tough break at life. He gave them a chance to find their purpose by making parts for wooden boats. The sheer coincidence that Essex Industries gave people like my sister a second chance felt like a sign from above.

"We make canoe paddles, seats, gunwales, thwarts, picnic tables, and woven baskets," he said, shouting to be heard over the din of power tools. "Top-of-the-line quality from some extraordinary people."

In the next room, a row of vises on a workbench sat closed, with protective earmuffs hanging from each thickness-adjustment crank. The bench was scarred from use and above it there were various drawers labeled with words such as *tapered drill bits*, *planer blades*, and *plug cutters*. I gradually became conscious of the fact that everything looked ancient. The machinery was dark with grease and made of heavy prewar cast iron. Every surface was coated in a fine layer of sawdust that gave the shop the appearance of a sunken ship viewed in grainy black-

and-white documentary footage. Each piece of machinery was supervised by a skilled employee who helped a disabled employee operate it safely. The supervisors had evaluated everyone's skills, and if a person couldn't operate a saw, for example, they sorted offcuts by shape and size, or packed orders.

"Everyone has a role here. Nobody gets left behind," Preston said.

We walked from the saw room to the sanding area, the stain booth, and the assembly room. Seats and thwarts at various stages of production were stored in overhead racks. About a dozen workers assembled canoe seats at a table piled high with spools of natural cane.

I swallowed hard against the emotions of seeing people with challenges like Lucy's, in a happy, vocational setting. "This is more than rehabilitation," Preston said. "It's taking care of people who can't fully take care of themselves."

At that moment I considered telling Preston that my sister had died exactly fifteen years and one week ago, and that I had thought of her every day since, and that I wished she could have worked in a place like this. I didn't let the words come out of my mouth, though. I didn't want him to think that I felt sorry for his workers, or that I had come here only because of Lucy's disability. I came because more than one professional boatbuilder told me Essex made the finest hardwood canoe parts in New York, period.

Preston led me into the milling area, where ash planks two or three feet wide and twenty or thirty feet long were stacked in the center of the room. He flipped the top piece onto the table saw and introduced me to an employee named Mike, who had a difficult time using the equipment because of his cerebral palsy. I reached out to shake Mike's hand but then realized his right arm was hanging limp by his side. He used his left arm to brace the ash plank against his hip, which jutted out from his body at an uncomfortable-looking angle. I skipped the handshake and patted him on the back instead.

"I never used a saw before I worked here," Mike said, laboring through his words.

Preston helped Mike push the ash board through the table saw, and then they examined the cut surface together. It was straight as can be and free of knots, an important trait for gunwales, which need to withstand an extraordinary amount of torque. Anytime the canoe passed over the crest of a wave, the gunwales would flex just enough for the hull to spring back into position in the trough of the wave. Preston asked me if I wanted to take the raw lumber home or if he should cut out the gunwales for me.

"Let us help you," he said.

"I would be honored."

I showed Preston the measurements for my canoe and he assisted Mike with cutting out my gunwales. First, they sent the ash through the power planer to take it down to size, then sent it across the jointer table, a special power tool that flattens wood surfaces. Mike's hands strained for every movement. He took his time concentrating on each surface. I couldn't have done what he was doing. It would have taken me weeks to learn how to rip-saw gunwales, and I might have broken another window or lost a finger in the process. I was so grateful for their help.

"They're perfect," I said, balancing the twenty-foot-long gunwales in my hands like there was something sacred about them, knowing that ash has a lot to teach the world about survival and adversity.

I stopped by the crafts room and picked out two hand-caned seats, one wooden paddle carved out of cherry, an ash thwart to span and stabilize the middle of the boat, and maple decks for the bow and the stern. Preston scheduled my gunwales to be delivered by freight courier. I was relieved to have the necessary trim parts in hand—parts that would have pushed me months over schedule had I made them myself. The young woman at the front desk handed me my new paddle and smiled.

"Will you send us pictures when you're done?" she asked.

"Yes. I promise."

On the drive home, I couldn't shake the feeling that somehow it was my fate to discover Essex Industries. What else could it be? I had never heard of them before, and yet total strangers pointed me in that direction and said, go, find what you need there, those nice people will help you finish your canoe—and they did. I found the wood from a dying species that soon wouldn't even exist on this earth. More importantly, I found people who talked and smiled like Lucy, who reminded me of the lightness of her being, her laugh, and all the reasons why I loved her. I got to watch them doing things with their hands that Lucy never could have done, and they looked happy and fulfilled doing it. I found it enormously gratifying to support their work by including it on my canoe. I had promised Lucy in the beginning that I would honor her with this build, and in a roundabout way that I never could have predicted, I was able to fulfill my promise. Maybe the discovery of Essex Industries was an early Christmas present from Dad. If it was, then it certainly would rival the toboggan for the best gift ever.

CHAPTER 31

FLOATING

Back home, I had two weeks to finish. My nearly completed canoe was a compelling sight, as if it had blossomed from out of nowhere, grown like a ponderosa sapling from the floor of my house itself. The ash gunwales slid over the sheerline strip like a glove and I fastened them to the hull with epoxy and screws. While those cured, I installed the triangular decks on each end, and the thwart amidships, bringing the sides of the hull into proper width. The hand-caned seats were bolted on last. All the wood species I had used were now fused together in a seamless way that paralleled the blending of different grape varieties, each one contributing part of itself to make a more complex wine.

For months, I had read about so-called traditional boatbuilding, and it struck me now that the more important question was "Whose tradition?" Canoes had always been a reference point to their place and time. North American tribes and fur traders used building materials available in their era, and likewise I wanted the canoe to reflect my unique place in the world. I brought Dad's rodeo belt buckle to a local marine supply shop, where I devised its nautical equivalent out of half-round strips of bronze and a bronze rope tie-down that resembled a bull's nose ring. I screwed these pieces to the bow stem. In boat parlance, they were called cutwaters, but to me, they were a bit of cowboy bling to honor my own tradition.

With one week left, my canoe was finished in the sense that all the parts and pieces of the tensioned wood spring had been installed. Before I could apply varnish, though, there was one final bit of woodworking to attend to: rounding off the square edges of the gunwales. For that task, I used a spokeshave, the final tool in Dad's toolbox that I hadn't yet touched. It had two metal handles that looked like bird's wings, with a cutting blade in the middle. Its purpose was to carve off thin fractions of wood, like a vegetable peeler shaving skin from a potato. Leaning my hip against the starboard hull, I propelled the tool with my whole body, reaching forward and pulling back like I was rowing a boat. The first long stroke of the spokeshave made a nice *zip* sound. With every twist of my hands to ease off the cuts, curlicue shavings fell to the floor. Over the course of the day, shavings piled up and filled the house with the distinctive burnt popcorn aroma of ash. Each time I pulled the blade over the wood, a wave of satisfaction passed through me. My body now seemed to know by instinct how to carve, as if I had mastered the art of delivering power to a small cutting edge. Dad's spokeshave became more than just an extension of my body and mind. It helped me see how over my whole life I had possessed the potential to learn these skills.

By late in the afternoon, I was sweaty from my exertions and had taken the canoe as far as I could go with Dad's tools. The surfaces only needed amalgamating with sandpaper before I applied varnish. What I had learned about sanding was that no amount of elbow grease could fix or change what I had already done. The deepest mistakes, the gouges, and scratches, would always be here, even if I could lessen their depth slightly. Pointing out my mistakes was one thing but forgiving myself for them was another. What would the ocean or the prairie be without deep swells? My mistakes added luster to the boat, and they made me change my ways from those early days when I didn't know what I was doing. I wondered whether I could've succeeded in the end if I hadn't failed at the beginning.

With sanding finished, I pried open a can of marine gloss varnish. The tawny syrup smelled delicious, almost edible, a mix of spice, pine, and turpentine solvents. The aroma took me back to childhood days helping Dad seal oak trim for the interior doors and windows of our house. I dipped a paintbrush into the varnish and let the viscous liquid drip back into the can. I brushed a thin coat over a small area of the hull, then dipped the brush into the can again and moved onto the next section. First this section, then that one, brushstroke by brushstroke, little and often, the boat glistened. As the varnish dried and hardened, it created a fragile and translucent yellow glow. The nuances of the wood and varnish shifted under different lighting and shadows. When I studied it from up close, I saw the wood grain; when I stood across the room, I saw the white glare of glossy reflections.

I worked on the finish over the next few days, hand-rubbing the hull with Scotch-Brite pads for hours, then applying another paper-thin coat of marine varnish, then rubbing the finish again and varnishing again. I could have added countless more coats, but a wooden boat would soak up every ounce of perfectionism I could throw at it. If I carried on obsessing over the varnish, I might never finish. At some point I just had to stop.

Or, in Dad's words, when he scolded me for filling my plate too high at the Faith Steakhouse's all-you-can-eat salad bar: "Plenty ought to be enough."

When my canoe gleamed like the still water of a cattle stock pond, when it seemed in its sleekness to be alive with the potential for floating, I knew it was done. The end result concealed the whole messy process. My misadventures building it had been smoothed over by the finished canoe's serene presence. It looked like it was intended to be that beautiful all along.

My house had transformed too—it had become Preszler Woodshop. In the dining room, on and above my workbench and the toolbox, I had placed drawers and cabinets holding drill bits, saw blades, glue, and screws. Sandpaper was everywhere, spilling out of boxes on the floor, filling the linen closet, and littering Caper's bed, where he chewed it to bits. The table saw sat against the chalkboard wall, which was filled with the drawings, sketches, and ramblings that came to me in the middle of the night and burned with urgency until I wrote them down. I had tacked up snapshots of employees and friends who visited over the last year, and their faces were now reflected in the mirrored surface of the hull. The small piece of petrified wood I had picked up in South Dakota sat next to the toolbox as a reminder of that Thanksgiving with my parents. At the time, of course, I had no idea it would be the last meal I would ever share with Dad, but since that was how it turned out, I was grateful that he told me to come home and bring a bottle of my fancy wine. I had saved that empty merlot bottle, too, and it sat on my workbench now, holding a dried rose.

I texted a photo of the finished canoe to my friends with one word: "Done!" My phone pinged with responses.

The first was from Dave: "Honey, it's beautiful. Like, museum quality."

And then Michael wrote: "It's truly a work of art, even if it floats. Your father would be proud of you, and so am I."

Allen texted back: "I see you grew into your shop clothes!"

Scott Roberts said: "Congrats, buddy, it turned out great! Let me know if you need more wood for the next boat. Nobody can build just one."

Peter said: "I thought you were full of shit, but you proved me wrong."

Preston sent his congratulations from the employees of Essex Industries.

Lastly, I heard from Mom, who was on her way to Peru to hike Machu Picchu. Turns out she decided to travel more, even without Dad.

"Your father could build anything," Mom said. "And now, so can you."

It seemed impossible that I wouldn't be building the canoe anymore, and that my house would no longer be littered with sandpaper, sawdust, varnish, fiberglass, and wood shavings. I'd finally be free to go about my days without Gilpatrick's golden chain of sequential construction hanging over my head and stalking my dreams. I'd be able to wear any article of clothing I wanted without checking for varnish stains and shaking out the sawdust. I'd clear off my workbench and eat meals there, sitting down as though it were a real dining table. I'd stop worrying about Caper chewing up offcuts or licking a toxic solvent behind my back. The thought of it all astounded me, but the journey had to end at some point.

In the wintry sunlight, the canoe somehow provided a perspective on the totality of my life that had been unclear until this point. There was my life before Dad died, and there was my life after his death, when all my navigational guideposts shifted and I followed a magnetic pulse that felt right in my bones, even if it made no sense in my head. During the emotional furnace of the past year, like Job in the Bible, I remained steadfast despite my awful circumstances. All my life, I never disputed that God existed or that an omniscient force created the world as a constantly changing place, but the church told me I didn't belong. I liked to believe that it was progress that Dad had in mind when he gave me his toolbox—his way of saying I was worthy of inclusion in everything, I just had to find my purpose.

God had asked Job, "Have you given the horse strength? Have you clothed his neck with thunder?" No, God, but I gave my canoe its strength and I clothed its hull with glass. I had to build it the hard way, Dad's way, the little and often way, but eventually, I created something that was meant to exist, something that might have existed before I even made it. Perhaps in the end I felt that way about the life I built for myself, too.

A world that might allow such a life of boatbuilding and woodworking—which had tormented me for the past year—had now opened its doors and welcomed me inside. The canoe spoke to me of that world. It said: *You belong here, whomever you made yourself into, that is who you always were.*

ON THE MORNING of launch day, the anniversary of Dad's death, I was idly scrolling through Facebook when a post by Cornell University's marine sciences department caught my attention. They had circulated a video of a minke whale that was spotted in Peconic Bay just minutes before. Whales use familiar landmarks and the earth's magnetic field to guide their migratory journeys from Canada to the Caribbean each winter, feeding on schools of bunker down the coast. The minke had apparently followed the deep channel from the Atlantic into the waters around Robins Island, directly in front of my house. I grabbed Dad's binoculars and raced outside to the bulkhead just in time to see the whale's enormous black form emerge on the surface. It exhaled through its blowhole in a giant, gaping, extraordinary torrent of air that blew mist into the sky before it dove down and disappeared. For a minute, I stood there with my heart racing. I was in awe. Did this minke visit the waters in front of my house last year, too, on the same day? It might have been a sign; it was *not* just another day.

A flock of ducks cupped their wings and dropped down onto the bay. They made silly sounds, cooing and quacking at each other. Some blast of arctic air had driven them south, guided by their internal clock and compass. The earliest canoeing explorers relied upon seabirds and whales to help them navigate the open sea, and I was finally ready to join them on the water. It was time to exchange the blades of Dad's tools for the blade of my canoe paddle.

I had to admit that December paddling conditions were not ideal, but I was determined to finish my journey and keep the promise I made one year ago. Peter and John helped me carry the canoe from the house to the beach, and then they departed. I needed to do this alone.

Dressed in winter gear, I slung Dad's rope over my shoulder. If disaster and oblivion had followed his rope over the decades from the rodeo arena to Vietnam and back, so too did family, as when Dad and

Chili towed Lucy and me around the prairie on our toboggan. In a sense, Dad's rope was an immortal part of our family, and I was ready to add my own experience to its history.

Under a somber gray sky with swirling clouds, I walked the canoe into a foot of icy water, soaking my work boots. I grabbed both gunwales and hopped in. The canoe wobbled back and forth while I gained my balance. The simple act of sitting in it made my chest burst with pride. I plunged my paddle into the water and twisted my body to pull it behind me. Three hard strokes and the canoe moved away from shore, gliding soundlessly over submerged jetties. I tracked east along the eddy line and seawater slapped against the hull, sending up icy spray that stung my eyes. Just outside the cut to Marratooka Point the tide ran out in slow brown coils that were laced with oak leaves and the occasional whiteness of a split tree branch bobbing under the surface. The salt marsh and seagrass beds spread out behind the beach like a low table, having been scrubbed clean by wind and waves. I gulped drafts of cold air and let my eyes go out of focus, watching the scene ahead of me turn into a soft blur of colors: the dark wintergreen *Spartina*, the lead-colored water, and the rusted-mahogany brilliance of my canoe.

I had paddled only a few yards into the deep channel when a light but cutting northeast wind ruffled the water. The paddling got a little harder. I moved the canoe slowly and erratically across the channel, into the wide expanse of Peconic Bay. In the swirling currents there was a constant renewal, a perpetual mixing: half seawater flowing west from the Atlantic, half bay water flowing east from inland tributaries. The waves of the Peconic were not huge but were steep and sharp-crested, and the wind tore off them at the top, leaving my paddle blade flailing at air one moment and sunk too deep the next. Frothy waves splashed over the deck and into the canoe. I was blown sideways, fighting the current, spray flying from the paddle with every release. The bow rose, bucking up the trough of one wave, shedding a string of water off the deck in droplets. When the bow fell, the gray sky enveloped the crest of another wave. I saw in action the design of the canoe's tensioned spring and all its parts supporting one another. The hull, held taut by the gunwales, straddled the waves with air underneath. I kept paddling, first this stroke, then the next one, until a curtain opened ahead of me. The infinite shades of blue on the sea were transformed when the sun broke through the clouds and unfurled a translucent ray of golden light.

With Robins Island in sight now, I paddled in long, precise strokes, setting a rhythm for myself, whipping the blade behind me and plunging it back down in front. I found my swing and surged through the waves. The beat went up a notch, and then another. I counted down the strokes, starting at ten, like I was bearing down on some imaginary finish line. The varnished hull was sleek and lithe, thrusting forward between my strokes like a galloping horse. God had told Job of his horse, "He devours the distance with fierceness." Yes, God, he did. My canoe moved with such strength and grace that I couldn't believe I had ever known a troubled moment building it.

Maybe a half hour had passed. Next to Robins Island, there was a narrow flume of water flowing over a shallow sandbar. With one final stroke I hurled the canoe into the current, which narrowed and squeezed into soft humps, haystacks, and standing waves. I got a sense of the canoe flying in three dimensions, in slow motion, and for the first time, I smiled. I was having fun, lots of it. I chuckled like I was that little farm boy again, riding on the toboggan with Lucy at Christmastime. I couldn't have known it thirty years ago, but our toboggan wasn't just wood and metal and twine, it was Dad's perfect idea of floating.

When I emerged from the chute on the sheltered leeward side of Robins Island, the environment underwent an abrupt change. I paddled into a cove and glided onto a glassy swell so wide it seemed to be a pond. A bald eagle perched on the limb of an oak tree jutting up from a cliff that was creased with streams of water flowing down its face like tears. The canoe reflected a violet color as it dipped and caught the late afternoon light. I nosed alongside the Robins Island dock and there, at last, I tethered the canoe to the dock with Dad's rope.

I had arrived. I did it. I felt like I'd cry, though I didn't. I only smiled. It seemed like such a little thing and such a big thing at once—like a secret I would tell myself the rest of my life, though I didn't understand the meaning of it all yet. I sat there for several minutes taking everything in, listening to the creaks and sighs of tiny ripples lapping against the hull. The cove near the dock was ringed by driftwood and boulders the size of cars—the disordered rocky remains of the Laurentide Ice Sheet. The North Fork was once a glacial wasteland. It was a wonder of the world that life had survived and evolved here in the old wreck and spill of rocks. But those rocks were also telling me something that I had paddled all this way to learn. The spinning of the earth had sent water east to west, sliding up sounds and bays, ebbing and flowing like clockwork, bringing in good and carrying out bad, little and often. Now, there was only the silence and stillness of the water: what the rocks turned into after healing.

I reclined on the stern and trailed both my hands over the gunwales, letting my fingertips dip into the water. I leaned my head back and closed my eyes against the sun. *Thank you*, Dad, I said over and over again. *Thank you*, not just for the toolbox, but also for everything I could feel gathering up inside me, for everything that building this canoe had taught me, and everything I couldn't yet know, though was already seared into my soul like a branding iron sears cattle hides.

So much was unknown to me, but I didn't have to know everything. It was enough to trust that what I did mattered, that I understood the canoe's meaning without yet being able to say precisely how, like all those confusing instructions from *Building a Strip Canoe* that had coursed through my mind the past year. I could trust my hands, knowing they built this canoe, and it was enough. It was my own sacred and mysterious life, manifested in a colorful floating quilt made of wood.

As I untethered Dad's rope from the dock and coiled it around my arm, I felt his presence there with me. He would always be with me, no matter what, embodied in the tools and the wood of this canoe. Through his death, Dad gave me a new life. He did, it was true. I could now rightfully call myself a craftsman and a boatbuilder who lived in communion with nature. My canoe was my freedom. With that understanding, some inexplicable pain deep inside me evaporated. In its place, I felt a flood of gratitude. I felt whole. I was my own man now—all of me—and I was ready to paddle home.

THAT NIGHT, I dried off the canoe as it rested on the floor of my living room. As much as I had wanted it, and as much as I understood what it meant to me, the canoe wasn't the most important thing I would take away from the experience. Immediately after the maiden voyage, even as I dragged the canoe onto the beach and opened the patio doors to let Caper run free, an expansive sense of calm had enveloped me. On the paddle home from Robins Island, I had thought about Mom and Dad, and all the people who have ever loved family heirlooms, looked out for them, sought them when they were lost, and tried to preserve them so they could be passed down to the next generation. And there had come a singular moment paddling home when I realized with startling clarity the meaning of my year as a boatbuilder. It unfolded inside me like a secret waiting to be told. Dad's lesson was not specifically about scraping paint off bricks, or growing grapes, or building boats. It was much more than that. It had everything to do with life itself.

I had started this journey believing that no matter how long I lived, Dad wouldn't have been able to give me the all-encompassing love that I craved. Maybe he was banking on that when he understood that he was going to die, because when the moment came for him to say goodbye, he didn't hold back a single ounce of his love. He gave me the biggest gift of all. It was a lot more than a banged-up toolbox and a taxidermied duck. It was everything he had.

Just as every wood shaving is a unique imprint of a tree's growth rings and the blade that cut it, so too had I been shaped and forged by a lifetime of little gestures that added up to something bigger. There were the days I had spent with Dad in his shop, or riding horses, or duck hunting, or tooling around country roads in Old Yeller, not speaking more than ten words to each other. There were the years Mom answered my phone calls at her office desk just to show up for me, and to listen. There were the moments I held Lucy's frail hands and stared into her eyes, knowing she couldn't utter a word back to me, but knowing she loved me, nonetheless. Theirs was a silent kind of love that ran deep and did not grow old, even in a culture that said otherwise. It was a love shown in actions, not words.

I took the long way home to find myself, and when I arrived, I learned that there was a greater love at the heart of parenting, and of being a good son. It was a love cultivated slowly as time marched on: a growing rather than diminishing acceptance of each other.

While it was true that I had built a boat, Dad had been working on his own project, something much bigger and more complicated than a canoe, something that took much longer than a year to complete.

I am the man my Dad made, little and often.

EPILOGUE

Three years later, I answered a phone call from the veterinarian.

"I'm sorry, we tried to save Caper, but it was too late," she said. "His heart was just too big."

I paddled my canoe halfway between the house and Robins Island, to the place where water flowed swiftly through the channel during the changing of the tides. There I scattered Caper's cremated ashes overboard and watched him float away to the far and boundless sea. It made sense to leave him here, where I could remember how his companionship had changed my life. The spring schools of bunker would swim in and out, the summertime scallops would flutter around the eelgrass, and the whales would migrate through, all carrying a bit of Caper with them.

I hauled the canoe back to the storage barn, which served as the new headquarters of Preszler Woodshop, a business I started the year after finishing my first canoe, and which I managed on the side in addition to running the winery. I had used the canoe often during the intervening years, but on that day, I hung it in the rafters and vowed never to paddle it again. It could not serve a higher purpose than helping me say goodbye to my best friend, the pup who was with me for all of it, who had even sat on Dad's lap while he watched football on Thanksgiving.

Twenty-two days after Caper died, the winery owner, Michael, died from cancer. I said goodbye to a great man who was both my boss and, in some ways, like a second father who had generously nurtured my career in the wine industry while embracing my newfound passion for wooden boats.

I began to realize that I couldn't live on the water forever, though I was reluctant to admit it at first. I wanted to buy a house, to own some land where I could plant an orchard and live close enough to the water to access it when I wanted to, but not so close that the sight of it every day reminded me of Dad and Caper. The realization needed time to sink in.

I walked down the beach, wondering what it would be like to leave. I loved living here and building my first canoe here changed my life for good. I could have forgotten everything I had done here during this season in the middle of my life, but I didn't forget how I had come to feel like the scrub oak and *Spartina*, sending down roots into the glacier's ruins and stretching limbs skyward.

The land and the sea where I had found myself were never really mine. Someone had been here long before me, and someone else long before them. Something important had happened here, a glacier or a nomadic tribe had perched on the bluff overlooking the bay, long before I had ever moved in and launched my canoe from that spot. This piece of earth, this beach, this bay, this sacred place would remain here without me, for the next generation, and the next. How peaceful it was to walk away and let it be.

ACKNOWLEDGMENTS

Few endeavors offer as much opportunity for common effort as the making of a book. With that in mind, I want to convey my deep appreciation to the following people.

Adam Chromy, my delightfully gutsy agent, who emailed out of the blue one day and asked if I'd ever considered writing a book. Of course, the rest is history. He has offered a rock-solid foundation of advice, edits, and encouragement, and has been indispensable to the story development on every conceivable level, large and small, from day one. As an added bonus, I became friends with Adam's wife, Jamie Brenner, a seasoned author who generously offered advice and talked me down off a ledge when the going got tough.

Mauro DiPreta, my brilliant editor at William Morrow, who took a chance on me after our first meeting and always followed through on his word. He is an absolute joy to work with, and I hope this is just the first of many books we write together. He taught me that the best editing is done with a whisper, along with the time to squeeze out one more draft. I am not sure how he did it, but he wielded the editorial scalpel so expertly that I hardly felt the pain and am eternally grateful for the cure. Most of all, I have to thank Mauro for the title of the book, which I recognized as a stroke of genius the instant he said it.

Liate Stehlik, Benjamin Steinberg, and Kelly Rudolph at the William Morrow imprint of HarperCollins, who marshalled the resources to acquire this little book and gave me a platform to share my story with the world. Ploy Siripant, who designed the stunningly beautiful cover and nailed it on the first try. And my team at William Morrow—Vedika Khanna, Alison Coolidge, and Amelia Wood—who took the baton when I got tired and finished the race for me.

Erwin Rosinberg, whose thoughtful reading of the manuscript, his many conversations with me about it, and his deeply insightful comments and suggestions improved it enormously.

Randee Daddona at *Newsday*, who was one of the first people to help me tell my story, and who wasn't satisfied just making a documentary film about me, she had to go out and win a New York Emmy for it, too.

David Cashion, my dear one, who travelled the back roads of South Dakota with me long ago, and who singlehandedly changed the trajectory of this project when he introduced me to Elizabeth Gilbert. And Liz, who I feel I have known in my heart since 1998, when David first bragged to me about his dear friend who had just published an article in *GQ*. . . . I could tell right away when I read her blurb that she understood every gut-wrenching word, and I felt seen.

Carrie Seim, a lifelong friend, author, journalist, comedian, Nebraskan, horse girl, Cyclone, and founder of Grandma Mojo's Moonshine Revival, who possesses a rare superpower for distilling grand ideas into a few words that might make a pithy chapter title.

Jenny Bower, who sketched the beautiful drawing of my canoe on the title page, and whose friendship and artistry inspire me every day to be a more earnest human.

Ninah Lynne and her late husband Michael, the owners of Bedell Cellars, whose encouragement, support, generosity, and passion have absolutely formed the bedrock foundation of my entire adult life and career. Also, the entire team at Bedell Cellars, who have made writing this memoir possible.

Richard Olsen-Harbich, a dear friend, fellow Cornellian, and the winemaker at Bedell Cellars, whose delicious fermentations were not only stars of the book but also helped me endure many late nights writing it.

Laina Albrecht, the best friend I could ever hope for, who possesses the kind of generosity that desires great things for her friends, and who has been my closest confidante from high school marching band to the coronavirus pandemic and beyond.

Randy Jones, who was by my side in the 1990s, when we were timid midwestern boys trying to make it in the big city, and whose emotional response to my first draft of this book gave me a sliver of hope that I might someday be able to help guys like us.

My Instagram followers and the extensive global network of boatbuilders and woodworkers I have met through that platform, who have supported me enthusiastically from the first day I posted a heavily filtered photo of my dog eating a ham sandwich. I will never be cynical about social media because they've made me feel part of something big, broad, creative, and thrilling. This includes the first people who read the finished manuscript: Isaac Mizrahi, Kevin O'Connor, and of course Nick Offerman, AKA Ron Swanson, an accomplished canoe builder in his own right, who graciously took the time to read my book and offer up a heartwarming blurb that may have left me misty-eyed at one point.

Finally, the bane of writing is self-doubt; the gift of writing is the real friends and family who show up to save you. I have felt all these people strongly in my corner at one point or another during the writing process: Sam Abrams, Uncle Norman Andenas, Elizabeth Andre, Dave Bastien, Ben Bentley, Carmindy Bowyer, Julie Buckles, Steve Carlson, Molly Deegan, Christina Delgado, Christian Dunbar, Angela Eissler, Peter Endriss, Christian Ercole, Jennifer Galvin, John Gidding, Genevieve Gorder, Peter Gundersen, Allen Haveson, Matthew

Hoffman, Christine Hoover, Gary Hoover, Brad Howarth, Elizabeth Hudson, Amy Finno Israel, Woozae Kim, Dave Mowers, Tim Pavlish, Charly Ray, Matt Rizzo, Jim Rosenthal, Amy Schiffman, Josh Swan, Brittany Kahan Ward, Steve Warren, Morgan Andenas Weber, and Aunt Cecelia Wittmayer. Their love, confidence, and continual support made writing a memoir possible in the first place. Without friends and family, there would be no books.